Agricultural Labour

Problem and Policy Implications

DR. S.S. CHHINA

REGAL PUBLICATIONS
New Delhi - 110 027

AGRICULTURAL LABOUR
Problem and Policy Implications

ISBN 978-81-8484-190-9

© 2012 S.S. CHHINA

All rights reserved with the Publisher, including the right to translate or to reproduce this book or parts thereof except for brief quotations in critical articles or reviews.

Typeset by
RAHUL COMPOSERS
Lakshmi Niwas, 760, Pocket-D, New Highway Apartments
Lok Nayak Puram, New Delhi - 110 041

Printed in India at
MAYUR ENTERPRISES
WZ Plot No. 3, Gujjar Market, Tihar Village, New Delhi - 110 018

Published by
REGAL PUBLICATIONS
F-159, Rajouri Garden, New Delhi - 110 027 • Phone : 45546396
E-mail : regalbookspub@yahoo.com

Contents

Preface

India is one of the largest agrarian economies of the world. Its agriculture sector is at the core of the economy's purchasing power. The agriculture produce sector is the most important component of the Indian commodity sector. India 's commodity sector comprises activities, regulations, institutions and producers, consumers, intermediaries, service producers and marketplaces that collectively cause and explain some part of the economy's total output. The agriculture produce sector comprehensively explains the geographic dispersion and statistical distribution of household incomes and purchasing power but the sector itself is under unprecedented stress. There is increasing pressure on all segments of the agriculture produce economy to respond to the challenges that the global markets pose in the new post-WTO world trade order.

As the earlier marketing system was developed in the context of a planned economy afresh approach to marketing is necessary in the atmosphere of liberalization and globalization. The National Agricultural Policy indicates the need for new demand driven marketing set up. The Expert Committee was therefore appointed to suggest ways and means to meet this new challenge through strengthening and developing agricultural marketing system in the country.

The National Agriculture Policy (NAP) envisages promotion of demand driven agriculture growth catering to domestic as well as export markets. As the efficient agricultural marketing system lies at the core of agriculture growth, the Government of India constituted an Expert Committee to review the present system and suggest ways and means to strengthen and develop it.

Government is promoting organized marketing of agricultural commodities in the country through a network of regulated markets. Most of the state Governments and Union Territories have enacted legislations to provide for development of agricultural produce market.

Regulated markets have helped in mitigating the market handicaps of producers/sellers at the wholesale assembling level. But, the rural periodic markets in general, and the tribal markets in particular, remained out of its developmental ambit. It was envisaged that physical markets with facilities and services would attract the farmers and the buyers creating competitive trade environment thereby offering best of the prices to the producers/sellers.

Although every effort has been made to offer the most authentic position on the subject, claiming hundred per cent accuracy will be too tall a claim. Any error, omission and suggestions for the improvement of the book brought to my notice shall be thankfully acknowledged and incorporated in the next edition.

DR. S.S. CHHINA

1

AGRICULTURE : MEANING AND ISSUES

INTRODUCTION

Agriculture refers to the production of food, feed, fiber and other goods by the systematic growing of plants, animals and other life forms. The broad field of agricultural research is called agricultural science. In modern usage, agriculture covers all activities essential to food/feed/fiber production, including all techniques for raising and "processing" livestock. Thus, it encompasses many subjects, including aquaculture, agronomy, animal husbandry, and horticulture. Each of these subjects can be further subdivided: for example, agronomy includes both organic farming and intensive farming, and animal husbandry includes ranching, herding, and intensive pig farming. Modern vertically-integrated agribusinesses play a major part in modern agriculture.

Agriculture has been a key component in the development of human civilization—in fact, it is widely believed that the domestication of plants and animals allowed humans to settle and give up their previous hunter-gatherer lifestyle. A great shift in agricultural practices has occurred in the past hundred years. The Haber-Bosch method for synthesizing nitrogen has supplemented the traditional practice of recycling nutrients with crop rotation and animal manure. Synthetic nitrogen, along with pesticides and mechanization, greatly increased the yields of agriculture in the early 20th century. Further yield increases were experienced later when high-yielding varieties of common staple grains such as rice, wheat, and corn were introduced as a part of the Green Revolution. The Green Revolution exported the technologies of the developed world out to the developing world. Thomas Malthus famously predicted that the Earth would not be able to support its exploding population, but technologies such as the Green Revolution have allowed the world to produce a surplus of food.

Some governments have subsidized agriculture to ensure an adequate food supply. These agricultural subsidies are often linked to the production of certain commodities such as wheat, corn, rice, soybeans, and milk. These subsidies have been noted as protectionist, inefficient, and environmentally damaging. Proponents of organic farming such as Sir Albert Howard argued in the early 1900s that the emphasis on pesticides and synthetic fertilizers damage the long-term fertility of the soil. While this feeling lay dormant for decades, as environmental awareness has recently increased there has been a movement towards sustainable agriculture by some farmers, consumers, and policymakers. One of the major forces behind this movement has been the European Union, which first certified organic food in 1991 and began reform of its Common Agricultural Policy (CAP) in 2005 to phase out commodity-linked farm subsidies, also known as decoupling.

As in 2006, an estimated 36 percent of the world's workers were employed in agriculture (down from 42% in 1996), making it by far the most common occupation. However, the relative significance of farming has dropped steadily since the beginning of industrialization, and in 2006—for the first time in history—the services sector overtook agriculture as the economic sector employing the most people worldwide. Also, agricultural production accounts for less than five percent of the gross world product (an aggregate of all gross domestic products).

Agriculture can be viewed on a spectrum dependent upon the intensity of the methods. At the one end lies the subsistence farmer who farms a small area with limited inputs and produces only enough food to meet the needs of his or her family. At the other end lies commercial intensive agriculture, including industrial agriculture. Industrial agriculture involves large fields and/or numbers of animals, high resource inputs (pesticides, fertilizers, etc.), and a high level of mechanization. These operations achieve economies of scale and require large amounts of capital in the form of land and machinery.

Agricultural production extends well beyond the traditional grains and includes fodder, (starch, sugar, alcohols and resins), fibers (cotton, wool, hemp, silk and flax), fuels (methane from biomass, ethanol, biodiesel), cut flowers, ornamental and nursery plants, tropical fish and birds for the pet trade, and both legal and illegal drugs (biopharmaceuticals, tobacco, marijuana, opium, cocaine).

The twentieth century saw changes in agricultural practice, particularly in agricultural chemistry and in mechanization. Agricultural chemistry includes the application of chemical fertilizer, chemical insecticides, and chemical fungicides, analysis of soil makeup and nutritional needs of farm animals.

Between 1950 and 1984 what is sometimes called the Green Revolution transformed agriculture around the globe and world grain production increased by 250%. Mechanization has enormously increased farm efficiency and productivity in most regions of the world, due especially to the tractor and various "gins" (short for "engine") such as the cotton gin, semi-automatic balers and threshers and, above all, the combine.

Other recent changes in agriculture include hydroponics, plant breeding, hybridization, gene manipulation, better management of soil nutrients, and improved weed control. Genetic engineering has yielded crops which have capabilities beyond those of naturally occurring plants, such as higher yields and disease resistance. Modified seeds germinate faster, and thus can be grown on an accelerated schedule. Genetic engineering of plants has proven controversial, particularly in the case of herbicide-resistant plants.

Engineers may develop plants for irrigation, drainage, conservation and sanitary engineering, particularly important in normally arid areas which rely upon constant irrigation, and on large scale farms.

The processing, packing and marketing of agricultural products are closely related activities also influenced by science. Methods of quick-freezing and dehydration have increased the markets for many farm products.

Animals, including horses, mules, oxen, camels, llamas, alpacas, and dogs, are often used to help cultivate fields, harvest crops, wrangle other animals, and transport farm products to buyers. Animal husbandry not only refers to the breeding and raising of animals for meat or to harvest animal products (like milk, eggs, or wool) on a continual basis, but also to the breeding and care of species for work and companionship.

Airplanes, helicopters, trucks, tractors, and combines are used in Western (and, increasingly, Eastern) agriculture for seeding, spraying operations for insect and disease control, harvesting, aerial topdressing and transporting perishable products. Radio and television disseminate vital weather reports and other information such as market reports that concern farmers. Computers have become an essential tool for farm management.

Ploughing rice paddies with water buffalo, in Indonesia.According to the National Academy of Engineering in the United States, agricultural mechanization is one of the 20 greatest engineering achievements of the 20th century. Early in the century, it took one American farmer to produce food for 2.5 people. By 1999, due to advances in agricultural technology, a single farmer could feed over 130 people.

In recent years, some aspects of intensive industrial agriculture have been the subject of increasing debate. The widening sphere of influence held by large seed and chemical companies, meat packers and food processors has been a source of concern both within the farming community and for the general public. Another issue is the type of feed given to some animals that can cause bovine spongiform encephalopathy in cattle. There has also been concern over the effect of intensive agriculture on the environment.

The patent protection given to companies that develop new types of seed using genetic engineering has allowed seed to be licensed to farmers in much the same way that computer software is licensed to users. This has changed the balance of power in favor of the seed companies, allowing them to dictate terms and conditions previously unheard of. The Indian activist and scientist Vandana Shiva argues that these companies are guilty of biopiracy.

Soil conservation and nutrient management have been important concerns since the 1950s, with the most advanced farmers taking a stewardship role with the land they use. However, increasing contamination of waterways and wetlands by nutrients like nitrogen and phosphorus are concerns that can only be addressed by "enlightenment" of farmers and/or far stricter law enforcement in many countries.

Increasing consumer awareness of agricultural issues has led to the rise of community-supported agriculture, local food movement, "Slow Food", and commercial organic farming.

HISTORY OF AGRICULTURE

Mehrgarh, one of the most important Neolithic (7000 BCE to 3200 BCE) sites in archaeology, lies on the "Kachi plain of Baluchistan, Pakistan, and is one of the earliest sites with evidence of farming (wheat and barley) and herding (cattle, sheep and goats) in South Asia."

Located near the Bolan Pass, to the west of the Indus River valley and between the present-day Pakistani cities of Quetta, Kalat and Sibi, Mehrgarh was discovered in 1974 by an archaeological team directed by French archaeologist Jean-François Jarrige, and was excavated continuously between 1974 and 1986. The earliest settlement at Mehrgarh—in the northeast corner of the 495-acre site—was a small farming village dated between 7000 BCE–5500 BCE.

ANCIENT ORIGINS

Developed independently by geographically distant populations, systematic agriculture first appeared in Southwest Asia in the Fertile Crescent, particularly in modern-day Iraq and Syria/Israel. Around 9500 BCE, proto-farmers began to select and cultivate food plants with desired characteristics. Though there is evidence of earlier sporadic use of wild cereals, it was not until after 9500 BCE that the eight so-called founder crops of agriculture appear: first emmer and einkorn wheat, then hulled barley, peas, lentils, bitter vetch, chick peas and flax.

By 7000 BCE, small-scale agriculture reached Egypt. From at least 7000 BCE the Indian subcontinent saw farming of wheat and barley, as attested by archaeological excavation at Mehrgarh in Balochistan. By 6000 BCE, mid-scale farming was entrenched on the banks of the Nile. About this time, agriculture was developed independently in the Far East, with rice, rather than wheat, as the primary crop. Chinese and Indonesian farmers went on to domesticate mung, soy, azuki and taro. To complement these new sources of carbohydrates, highly organized net fishing of rivers, lakes and ocean shores in these areas brought in great volumes of essential protein. Collectively, these new methods of farming and fishing inaugurated a human population boom dwarfing all previous expansions, and is one that continues today.

By 5000 BCE, the Sumerians had developed core agricultural techniques including large scale intensive cultivation of land, mono-cropping, organized irrigation, and use of a specialized labour force, particularly along the waterway now known as the Shatt al-Arab, from its Persian Gulf delta to the confluence of the Tigris and Euphrates. Domestication of wild aurochs and mouflon into cattle and sheep, respectively, ushered in the large-scale use of animals for food/fiber and as beasts of burden. The shepherd joined the farmer as an essential provider for sedentary and semi-nomadic societies.

Maize, manioc, and arrowroot were first domesticated in the Americas as far back as 5200 BCE. The potato, tomato, pepper, squash, several varieties of bean, Canna, tobacco and several other plants were also developed in the New World, as was extensive terracing of steep hillsides in much of Andean South America.

In later years, the Greeks and Romans built on techniques pioneered by the Sumerians but made few fundamentally new advances. The Greeks and Macedonians struggled with very poor soils, yet managed to become dominant societies for years. The Romans were noted for an emphasis on the cultivation of crops for trade.

AGRICULTURE IN THE MIDDLE AGES

During the Middle Ages, Muslim farmers in North Africa and the Near East developed and disseminated agricultural technologies including irrigation systems based on hydraulic and hydrostatic principles, the use of machines such as norias, and the use of water raising machines, dams, and reservoirs. They also wrote location-specific farming manuals, and were instrumental in the wider adoption of crops including sugar cane, rice, citrus fruit, apricots, cotton, artichokes, aubergines, and saffron. Muslims also brought lemons, oranges, cotton, almonds, figs and sub-tropical crops such as bananas to Spain.

ENVIRONMENTAL IMPACT

Agriculture may often cause environmental problems because it changes natural environments and produces harmful by-products. Some of the negative effects are:

- (i) Loss of biodiversity
- (ii) Surplus of nitrogen and phosphorus in rivers and lakes
- (iii) Detrimental effects of herbicides, fungicides, insecticides, and other biocides
- (iv) Conversion of natural ecosystems of all types into arable land
- (v) Consolidation of diverse biomass into a few species
- (vi) Soil erosion
- (vii) Depletion of minerals in the soil
- (viii) Particulate matter, including ammonia and ammonium off-gassing from animal waste contributing to air pollution
- (ix) Weed Science-feral plants and animals
- (x) Odor from agricultural waste
- (xi) Soil salination

According to the United Nations, the livestock sector (primarily cows, chickens, and pigs) emerges as one of the top two or three most significant contributors to our most serious environmental problems, at every scale from local to global. Livestock production occupies 70% of all land used for agriculture, or 30% of the land surface of the planet. It is one of the largest sources of greenhouse gases—responsible for 18% of the world's greenhouse gas emissions as measured in CO_2 equivalents. By comparison, all transportation emits 13.5% of the CO_2. It produces 65% of human-related nitrous oxide (which has 296 times the global warming potential of CO_2) and 37% of all human-induced methane (which is 23 times as warming as CO_2). It also generates 64% of the ammonia, which contributes to acid rain and acidification of ecosystems.

PETROLEUM PRICE, AVAILABILITY, AND THE EFFECT ON AGRICULTURE

Since the 1940s, agriculture has dramatically increased its productivity, due largely to the use of petrochemical derived pesticides, fertilizers, and increased mechanization. Called the Green Revolution, this has allowed world population to grow more than double over the last 50 years. Some arguing that because every joule in modern food that one eats requires 5-15 joules to produce and deliver, it is inevitable that a decreasing supply of oil will cause industrial agriculture to collapse. Such a collapse would lead to a drastic decline in food production, food shortages and possibly even mass starvation, unless the uses of fossil fuels in agriculture can be efficiently replaced with alternatives. For example, by far the biggest fossil fuel input to agriculture is the use of natural gas as a hydrogen source for the Haber-Bosch fertilizer-creation process. Natural gas is used because it is the cheapest currently available source of hydrogen. Were natural gas to become too expensive, other sources (such as electrolysis powered by solar energy or hydropower) would have to be used to provide the hydrogen to create these fertilizers without relying on fossil fuels.

Oil shortages may force a return to organic agriculture methods. While some farmers using modern organic-farming methods have reported yields as high as those available from conventional farming (but without the use of fossil-fuel-intensive artificial fertilizers or pesticides), this may be more labor-intensive and require a population shift from urban to rural areas, reversing the trend towards urbanization which has predominated in industrial societies.

Besides adjusting for increased fuel costs, farmers are now planting non-food crops such as corn to help mitigate peak oil, with the result of lower food production. Others point out that rising food and fuel costs will limit the abilities of charitable donors to send food aid to starving populations.In the UN, some warn that the recent 60% rise in wheat prices could cause "serious social unrest in developing countries."

POLICY

Agricultural policy focuses on the goals and methods of agricultural production. At the policy level, common goals of agriculture include:

(i) Food safety: Ensuring that the food supply is free of contamination.
(ii) Food security: Ensuring that the food supply meets the population's needs.
(iii) Food quality: Ensuring that the food supply is of a consistent and known quality.
(iv) Poverty Reduction
(v) Conservation
(vi) Environmental impact
(vii) Economic stability

AGRICULTURE SAFETY AND HEALTH

Satellite image of circular crop fields characteristic of center pivot irrigation in Haskell County, Kansas in late June 2001. Healthy, growing crops are green. Corn is growing leafy stalks, but Sorghum, which resembles corn, grows more slowly and is much smaller and therefore paler. Wheat is a brilliant gold as harvest occurs in June. Brown fields have been recently harvested and plowed under or lie fallow for the year in United States.

Agriculture ranks among the most hazardous industries. Farmers are at high risk for fatal and nonfatal injuries, work-related lung diseases, noise-induced hearing loss, skin diseases, and certain cancers associated with chemical use and prolonged sun exposure. Farming is one of the few industries in which the families (who often share the work and live on the premises) are also at risk for injuries, illness, and death.

On an average year, 516 workers die doing farm work in the U.S. (1992-2005). Of these deaths, 101 are caused by tractor overturns.

Every day, about 243 agricultural workers suffer lost-work-time injuries, and about 5% of these result in permanent impairment.

YOUNG WORKERS

Agriculture is the most dangerous industry for young workers, accounting for 42% of all work-related fatalities of young workers in the U.S. between 1992 and 2000. Unlike other industries, half the young victims in agriculture were under age of 15.

For young agricultural workers aged 15-17, the risk of fatal injury is four times the risk for young workers in other workplaces. Agricultural work exposes young workers to safety hazards such as machinery, confined spaces, work at elevations, and work around livestock.

An estimated 1.26 million children and adolescents under 20 years of age resided on farms in 2004, with about 699,000 of these youth performing work on the farms. In addition to the youth who live on farms, an additional 337,000 children and adolescents were hired to work on U.S. farms in 2004.

On an average, 103 children were killed annually on farms (1990-1996). Approximately 40 percent of these deaths were work-related.

In 2004, an estimated 27,600 children and adolescents were injured on farms; 8,100 of these injuries were due to farm work.

2

Agricultural Policy and Economic Development Strategies

INTRODUCTION

Agricultural policy describes a set of laws relating to domestic agriculture and imports of foreign agricultural products and exports from home country. Governments usually implement agricultural policies with the goal of achieving a specific outcome in the domestic agricultural product markets. Outcomes can range from guaranteed supply level, price stability, product quality, product selection, land use or employment.

AGRICULTURE POLICY CONCERNS

An example of the breadth and types of agriculture policy concerns can be found in the Australian Bureau of Agricultural and Resource Economics article Agricultural Economies of Australia and New Zealand which says that the major challenges and issues faced by their agriculture industry are:

(i) marketing challenges and consumer tastes;
(ii) international trading environment (world market conditions, barriers to trade, quarantine and technical

barriers, maintenance of global competitiveness and market image, and management of biosecurity issues affecting imports and the disease status of exports);

(iii) biosecurity (pests and diseases such as bovine spongiform encephalopathy (BSE), avian influenza, foot and mouth disease, citrus canker, and sugarcane smut);

(iv) infrastructure (such as transport, ports, telecommunications, energy and irrigation facilities);

(v) management skills and labour supply (With increasing requirements for business planning, enhanced market awareness, the use of modern technology such as computers and global positioning systems and better agronomic management, modern farm managers will need to become increasingly skilled. Examples: training of skilled workers, the development of labour hire systems that provide continuity of work in industries with strong seasonal peaks, modern communication tools, investigating market opportunities, researching customer requirements, business planning including financial management, researching the latest farming techniques, risk management skills);

(vi) coordination (a more consistent national strategic agenda for agricultural research and development; more active involvement of research investors in collaboration with research providers developing programs of work; greater coordination of research activities across industries, research organisations and issues; and investment in human capital to ensure a skilled pool of research personnel in the future);

(vii) technology (research, adoption, productivity, genetically modified (GM) crops, investments);

(viii) water (access rights, water trade, providing water for environmental outcomes, assignment of risk in response to reallocation of water from consumptive to environmental use, accounting for the sourcing and allocation of water); and

(ix) resource access issues (management of native vegetation, the protection and enhancement of biodiversity, sustainability of productive agricultural resources, landholder responsibilities).

POVERTY REDUCTION

Agriculture remains the largest single contributor to the livelihoods of the 75% of the world's poor who live in rural areas. Encouraging agricultural growth is therefore an important aspect of agricultural policy in the developing world. In addition, a recent Natural Resource Perspective paper by the Overseas Development Institute found that good infrastructure, education and effective information services in rural areas were necessary to improve the chances of making agriculture work for the poor.

BASIC POLICY TOOLS

SUBSIDIES

Governments pay subsidies to encourage domestic production of a good. Such subsidies are necessary when a government wants to alter the free-market outcome in product markets. The subsidies transfer some costs of production from producers to the government, allowing production at above-market costs. The effect of Western food subsidies 360$ billion is overwhelmingly to reduce world prices. For decades the 3rd world has earned only half the free market price for food, its primary product. Consumption subsidies can also be used to alter markets. A consumption subsidy offsets a portion of the purchase price of a product.

PRICE CONTROLS

Price floors or price ceilings set a minimum or maximum price for a product. Price controls alter free-market outcomes by encouraging over-production by a price floor or over-consumption by a price ceiling.

IMPORT BARRIERS

A government can erect trade barriers to limit the quantity of goods imported (in the case of a Quota Share) or enact tariffs to raise the domestic price of imported products. These barriers give preference to domestic producers.

OBJECTIVES OF MARKET INTERVENTION

NATIONAL SECURITY

Some argue that nations have an interest in assuring there is sufficient domestic production capability to meet domestic needs in the event of a global supply disruption. Significant dependence on foreign food producers makes a country strategically vulnerable in the event of war, blockade or embargo. Maintaining adequate domestic capability allows for food self-sufficiency that lessens the risk of supply shocks due to geopolitical events. Agricultural policies may be used to support domestic producers as they gain domestic and international market share. This may be a short term way of encouraging an industry until it is large enough to thrive without aid. Or it may be an ongoing subsidy designed to allow a product to compete with or undercut foreign competition. This may produce a net gain for a government despite the cost of interventions because it allows a country to build up an export industry or reduce imports.

ENVIRONMENTAL PROTECTION AND LAND MANAGEMENT

Farm or undeveloped land composes the majority of land in most countries. Policies may encourage some land uses rather than others in the interest of protecting the environment. For instance, subsidies may be given for particular farming methods, forestation, land clearance, or pollution abatement.

RURAL POVERTY AND UNEMPLOYMENT RELIEF

Subsidising farming may encourage people to remain on the land and obtain some income. This might be relevant to a third world country with many peasant farmers, but it may also be a consideration to more developed countries such as Poland. They have a very high unemployment rate, much farmland and retains a large rural population growing food for their own use. Price controls may be used to assist poor citizens. Many countries have used this method of welfare support as it delivers cheap food to the poorest without the need to assess people to give them financial aid.

ORGANIC FARMING ASSISTANCE

Welfare economics theory holds that sometimes private activities can impose social costs upon others. Industrial agriculture is widely considered to impose social costs through pesticide pollution and nitrate pollution. Further, agriculture uses large amounts of water, a scarce resource. Some economists argue that taxes should be levied on agriculture, or that organic agriculture, which uses little pesticides and experiences relatively little nitrate runoff, should be encouraged with subsidies. In the United States, 65% of the approximately $16.5 billion in annual subsidies went to the top 10% of farmers in 2002 because subsidies are linked to certain commodities. On the other hand, organic farming received $5 million for help in certification and $15 million for research over a 5-year time period.

Fair trade rules to ensure that poor farmers in underdeveloped nations that produce crops primarily for export are not exploited to put local farmers in developing nations out of work - which advocates consider a dangerous "race to the bottom" in agricultural labour and safety standards. Opponents point out that most agriculture in developed nations is produced by industrial corporations (agribusiness) which are hardly deserving of sympathy, and that the alternative to exploitation is poverty.

ARGUMENTS AGAINST MARKET INTERVENTION

DUMPING OF AGRICULTURAL SURPLUSES IS HARMFUL TO DEVELOPING WORLD FARMERS

When rich countries subsidize domestic production, excess output is often given to the developing world as foreign aid. This process eliminates the domestic market for agricultural products in the developing world, because the products can be obtained for free from western aid agencies. In developing nations where these effects are most severe, small farmers could no longer afford basic inputs and were forced to sell their land.

"Consider a farmer in Ghana who used to be able to make a living growing rice. Several years ago, Ghana was able to feed and export their surplus. Now, it imports rice. From where? Developed countries. Why? Because it's cheaper. Even if it costs the rice producer in the developed world much more to produce the rice, he doesn't have to make a profit from his crop. The government pays him to grow it, so he can sell it more cheaply to Ghana than the farmer in Ghana can. And that farmer in Ghana? He can't feed his family anymore."(Lyle Vanclief, Former Canadian Minister of Agriculture [1997-2003])

According to The Institute for Agriculture and Trade Policy, corn, soybeans, cotton, wheat and rice are sold below the cost of production, or dumped. Dumping rates are approximately forty percent for wheat, between twenty-five and thirty percent for corn (maize), approximately thirty percent for soybeans, fifty-seven percent for cotton, and approximately twenty percent for rice. For example, wheat is sold for forty percent below cost.

According to Oxfam, "If developed nations eliminated subsidy programs, the export value of agriculture in lesser developed nations would increase by 24 %, plus a further 5.5 % from tariff equilibrium. ... exporters can offer US surpluses for sale at prices around half the cost of production; destroying local agriculture and creating a captive market in the process."

Free trade advocates desire the elimination of all market distorting mechanisms (subsidies, tariffs, regulations) and argue that, as with free trade in all areas, this will result in aggregate benefit for all. This position is particularly popular in competitive agricultural exporting nations in both the developed and developing world, some of whom have banded together in the Cairns Group lobby. Canada's Department of Agriculture estimates that developing nations would benefit by about $4 billion annually if subsidies in the developed world were halved.

AGRICULTURAL INDEPENDENCE

Many countries don't grow enough food to feed their own populations. These nations must buy food from other countries. Lower prices and free food save the lives of millions of starving people, despite the drop in food sales of the local farmers.

A developing nation could use new improved farming methods to grow more food, with the ultimate goal of feeding their nation without outside help. New greenhouse methods, hydroponics, fertilizers, R/O Water Processors, hybrid crops, fast-growing hybrid trees for quick shade, interior temperature control, greenhouse or tent insulation, autonomous building gardens, sun lamps, mylar, fans, and other cheap tech can be used to grow crops on previously unarable land, such as rocky, mountainous, desert, and even Arctic lands. More food can be grown, reducing dependency on other countries for food.

Replacement crops can also make nations agriculturally independent. Sugar, for example, comes from sugar cane imported from Polynesia. Instead of buying the sugar from Polynesia, a nation can make sugar from sugar beets, maple sap, or sweetener from stevia plant, keeping the profits circulating within the nation's economy. Paper and clothes can be made of hemp instead of trees and cotton. Tropical foods won't grow in many places in Europe, but they will grow in insulated greenhouses or tents in Europe. Soybean plant cellulose can replace plastic (made from oil). Lemon oil can replace car oil for lubrication. Ethanol from farm waste or hempseed oil can replace gasoline. Rainforest medicine plants grown locally can replace many imported medicines. This is why Thomas Jefferson said that the best thing you can do for your country is to grow a new crop species. Alternates of cash crops, like sugar and oil replacements, allow the farmer to make more money on the real market, reducing the farmer's dependency on subsidies in both developed and developing nations.

HIGHER MARKET PRICES

The cost to consumers for agricultural products is increased, either via hidden wealth transfers via the government, or increased prices at the consumer level, such as for sugar and peanuts in the US. This has led to market distortions, such as food processors using high fructose corn syrup as a replacement for sugar. High fructose corn syrup may be an unhealthy food additive, and, were sugar prices not inflated by government fiat, sugar would be preferred over high fructose corn syrup in the marketplace.

AGRICULTURAL AND ECONOMIC DEVELOPMENT STRATEGIES

By any measure, China and—more recently—India are striking economic success stories. A few decades ago, both countries were clearly among the world's poorest countries; now they are among the world's fastest-growing economies and are responsible for nearly all the recent global progress in poverty reduction.

In 1978 per capita gross domestic product (GDP) in India was $1,255--lower than the average for Sub-Saharan Africa, which stood at $1,757.1 Since then it has climbed steadily upward, reaching $2,732 in 2003. Even more spectacularly, China's GDP per capita, which stood at $1,071 in 1978, jumped to $4,726 in 2003. China's GDP per capita growth rate is almost double that of India. Moreover, the share of rural poor people fell from 33 percent in 1978 to 3 percent in 2001, according to official sources, or to around 11 percent, based on a poverty line of less than a dollar a day, according to World Bank estimates of 1998. Despite ongoing controversies regarding measures of poverty in China, both benchmarks depict an extraordinary decline in the incidence of poverty. India also achieved a downward trend in poverty, although the outcomes were not as dazzling as in China. According to official estimates, rural poverty in India dropped from 50 percent in 1979/80 to 27 percent in 1999–2000, the latest year for which data are available. Together these two countries accounted for a substantial drop in global poverty levels, from 29.6 percent of the world's population in 1990 to 23.2 percent in 1999.

Less well known than their recent blistering economic performance, however, is the role that agriculture has played in the transformation of these still heavily rural and agricultural countries. In China agricultural reforms were the starting point for economic liberalization--in other words, reforms began in the sector where the majority of poor lived, and they were largely the beneficiaries of reform--whereas in India reforms started with macroeconomic adjustment and trade and industrial policy, areas that did not benefit most of the poor. Although agricultural growth in India rose to more than 4 percent a year in the years immediately following the reforms (1992–96), it could not be sustained, and it slumped to about 2 percent a year during the period 1997–2003, severely affecting its contribution to economic growth and poverty reduction. The full potential of agriculture in India has yet to be unleashed. Now, in 2005, agriculture is once again high on the agenda of the Indian government, which wants to give a rural orientation to the entire reform and growth process. The reform experiences of China and India--similar in some ways and different in others--shed light on the enormous potential for investments and policies in support of pro-poor agricultural and rural growth to fight poverty and malnutrition in developing countries.

REFORMS IN CHINA AND INDIA

Reforms that directly strengthened agriculture were a major factor in China's economic growth and poverty reduction. Between 1978 and 1989, China underwent two distinct phases of agricultural reform, which first decentralized agricultural production through the household responsibility system, giving farmers much more leeway to decide what and how much to grow, and then liberalized the systems for pricing and marketing agricultural goods. Reported agricultural production growth immediately shot up, from 2.6 percent a year during 1966–76 to 7.1 percent a year during 1978–84. Furthermore, growth in agricultural productivity went from almost zero to 6.1 percent a year. Although production growth fell back to 2.7 percent a year during 1985–89 because of rising input prices, further reforms in the 1990s again raised production growth to 3.8 percent a year during 1990–97. As a result of the dramatic growth in agriculture, rural people found their incomes rising by 15 percent a year between 1978 and 1984.

But perhaps one of the most striking results of China's agricultural reforms was that they led to the creation of a whole new economic sector that became the most dynamic in China's economy: the rural nonfarm sector--the small-scale food-processing plants, machinery repair shops, and increasingly more modern and technology intensive industries that cropped up to meet growing demand among increasingly well-off farmers and to employ the millions of people whose labour was no longer needed on farms. Indeed, the whole structure of China's economy shifted. Whereas agriculture provided more than half of the country's GDP in 1952, it fell to 14 percent in 2004. Over the same period, the rural nonfarm sector went from providing almost none of GDP to more than one-third. The growth of this sector not only played a large role in reducing rural poverty in China, but also put pressure to reform on the urban sector, which has been the main engine of growth since the 1990s.

The story of agriculture in India is somewhat different. During the 1960s and 1970s, the Green Revolution, in which Indian farmers adopted new high-yielding varieties of wheat and rice, led to dramatic leaps in agricultural production and raised farmers' incomes. As a result, rural poverty fell from 64 percent in 1967 to 56 percent in 1973 and to 50 percent in 1979/80. Production gains from Green Revolution technologies continued through the mid-1980s and then slowed sharply. In the 1970s India had adopted subsidies for agricultural inputs, such as fertilizers and electricity for pumping irrigation water, and these subsidies grew to help maintain agricultural production but started placing a strain on government budgets.

Beginning in 1991 India instituted a series of sweeping macroeconomic reforms. Although these initial reforms were not directed toward agriculture, they helped stimulate a rise in agricultural growth by generating greater demand for a wide range of agricultural products and by leading to increased private investment in agriculture. From 1991/92 to 1996/97, agriculture grew at an annual rate of 4.1 percent and rural poverty fell only from 39.1 percent in 1987/88 to 37.3 percent in 1993, and further to 27.1 percent in 1999/2000. After the government opened the agricultural sector to international trade in the face of falling world prices of most agricultural products during the late 1990s, agricultural growth slowed again, averaging 2 percent between 1997/98 and 2003/04. Various studies have shown that whenever there is higher agricultural growth, there is greater poverty reduction in rural areas.

Now further steps are needed in India to again stimulate strong agricultural growth, including investments in roads, irrigation, and other infrastructure, improvements in education, and greater emphasis on growing high-value agricultural goods like fruits and vegetables instead of only cereals.

LESSONS FROM CHINA AND INDIA

What can we learn from the process of economic reform in these two countries? Does the sequencing of reforms matter? What lessons do the experiences of China and India offer for other developing countries and countries in economic transition? What could China and India learn from their own as well as each other's experiences?

To Reduce Poverty Faster, Begin with Agricultural Reforms

China's reforms led to acceleration in agricultural growth from 1978 to 2002 (4.6 percent per year, as opposed to 2.5

percent per year from 1966 to 1977). The most substantial decline occurred in the first phase of reform, from 1978 to 1984, when agricultural GDP jumped to 7.1 percent per year and the percentage of rural poor dropped from 33 to 11 percent of the population.

By launching market-oriented reforms in agriculture, China was able to ensure that economic gains were widespread and thus build consensus for the continuation of reforms. Besides, prosperity in agriculture favoured the development of rural nonfarm activities, which, by providing additional sources of income beyond farming, were one of the main factors behind China's rapid poverty reduction after 1985. As the rural nonfarm enterprises became more competitive than the state-owned enterprises, the government expanded the scope of policy changes and put pressure on the urban economy to reform. Reforms of the state-owned enterprises in turn triggered macroeconomic reforms, opening up the economy further.

In India, on the other hand, even though overall economic growth was high, it is clear that slower growth in agriculture was the major reason behind the slower poverty reduction. Prompted by macroeconomic imbalances, India's reforms began with macroeconomic and nonagricultural policy changes. The reforms led to impressive rates of economic growth in the 1990s, but since reforms were largely focused on the nonagricultural sectors, they had limited impact on poverty reduction. Agricultural policy changes occurred only at later stages, and even then were only partial. Therefore, the evidence suggests that successful agriculture-led reforms reduce poverty faster.

Make Reforms Gradually and Carefully

At the outset of reforms in China, policymakers withdrew central planning and reduced the scope of government procurement while expanding the role of private trade and markets. Thus they first created the incentives and institutions required by the market economy; then, in the mid-1980s, they began to open up markets. Studies show that the incentive reforms--in the form of greater land use rights, decentralized agricultural production management through the household responsibility system, and rises in procurement prices--from 1978 to 1984 had a greater impact on growth than did market liberalization reforms per se after 1984. Incentive reforms in China allowed markets to emerge gradually, so unlike other countries in transition, China did not experience a sudden collapse of central planning in the absence of market-based allocative mechanisms. Parallel with reforms in output markets, reforms in the pricing and marketing of inputs, including fertilizer, machinery, fuel, feed, seeds, and energy, have transformed a system of state-controlled quotas and prices into a largely market-driven system. Today the role of government is limited to quality control of input supplies. Subsidies for fertilizer and machinery imports and domestic manufacturing have also been eliminated. In the irrigation sector, the state is still responsible for large-scale investment, but farmers or local governments are responsible for local investments and maintenance of the lower end of the system.

This favourable sequence of reforms came about not so much through the planning of Chinese policymakers, but rather through their trial-and-error approach to reform. Instead of following a predetermined blueprint, they adopted new measures through experimentation—in the words of Deng Xiaoping, "crossing the river while feeling the rocks." Each new policy was field-tested and determined to be successful in selected pilot districts before the policy was applied nationwide and the next measure introduced. This gradual approach to reforms, beginning with the strengthening of market institutions and incentives and moving toward the opening up of markets, appears to lead to more substantial rates of growth and poverty reduction.

India's quite different experience also supports this assertion. India's reforms in the agriculture sector began with agricultural trade reforms, despite the fact that the incentive structure of Indian agriculture was highly distorted; the sector was, and still is, burdened with excessive regulations on private trading and most market activities. The liberalization of agricultural trade policies in the mid-1990s, coming before incentive and market reforms in the domestic arena, created a series of imbalances. Lowered protection against a backdrop of low international prices increased agricultural stocks in the late 1990s and led to an unprecedented accumulation of foodgrain stocks at home.

Reform Incentives before Opening Markets

China's experience with marketing reforms can be valuable for other economies transitioning from a centrally planned to a market system. Policymakers embarking on the reform path should first increase incentives for production and build the institutions needed to operate efficiently in a market economy before rushing to open up markets.

In a situation of food oversupply and liberalization of agricultural trade, farm support policies geared toward self-sufficiency lose their original rationale. In India minimum support prices and input subsidies, initially intended to encourage the adoption of new technologies and fuel agricultural growth, increasingly turned from incentives into inefficient and costly income-support interventions. It is clear that once support measures have completed their function, they need to be abolished. Otherwise they lead over time to inefficiencies and the crystallization of vested interests, resulting in the slowing of growth and poverty reduction.

China could learn from the experience of India and seek to encourage agricultural growth in the future while at the same time avoiding the large inefficient subsidies provided to its agricultural sector. This issue is of increasing relevance given the recent introduction of the direct transfer program to farmers and the emphasis placed by many scholars and government officials on increasing government support to agriculture and rural areas.

Although agricultural marketing reforms in India were limited, state governments were reluctant to implement them and thus their impact was reduced. In addition, a host of outdated domestic regulations under the Essential Commodities Act of 1955 continue to weaken the environment for agribusiness and private sector involvement in agricultural marketing, which could boost employment and efficiency. Against the backdrop of rising and diversifying food demand and liberalized agricultural trade, reform of these regulations is increasingly critical, as it has a direct impact on the capacity of the sector to adjust to the changing context.

Given that smallholder agriculture is predominant in both countries, farmers could be excessively penalized because they do not possess sufficient capital and information to manage the risks inherent in agricultural activities. While China and India are reconsidering current forms of agricultural and input subsidies, they should put in place well-targeted and innovative, cost-effective crop insurance policies to protect vulnerable farmers from drastic supply and price shocks.

One other important area is the strengthening of the network of support services for small farmers related to information, credit, and extension. India seems to be better off than China in these areas, particularly with regard to the institutional infrastructure of rural credit and marketing, although the reach of its services may not be perfect. The Indian experience shows that smallholder agriculture needs strong institutional support in these areas to grow and prosper.

In terms of trade liberalization, both countries made progress in reducing protection levels, but the weighted average tariff in India, at 29 percent, is almost double China's 16 percent. India has been able to sustain its current growth rate with lower inflows of foreign direct investment and a weaker export orientation than China. If India is to attain the target of 8 percent growth in GDP, it may do well to follow through with reforms to foreign direct investment in view of their potential to transfer know-how, managerial skills, and new technologies. China can offer valuable lessons in this regard.

The inevitable restructuring and adjustments involved in opening up agricultural trade flows will produce both winners and losers. Domestic producers of crops for which the country lacks a comparative advantage (such as edible oils in India and wheat and maize in China) are likely to suffer increasingly from falling prices induced by an increase in imports. In addition, broad-based structural adjustments in the economy may depress rural incomes and increase opportunities in the manufacturing and service sectors, located primarily in urban areas. These intersectoral adjustments are likely to result in a reduction in the size of the primary sector, which will release additional unskilled labour into the labour markets. The rural population will gain if it is able to shift to more profitable off-farm occupations. Investment in rural education will be crucial in increasing farmers' ability to move out of farming. It will also be important to increase investments in rural R&D and infrastructure in order to enhance productivity.

Membership in the World Trade Organization (WTO) can provide useful external pressure to improve efficiency and implement reforms, particularly for tradable inputs such as seeds, fertilizers, farm machinery, and pesticides, where markets are regarded as inefficient because of either government intervention or lack of infrastructure. The implementation of the various agreements under the WTO can facilitate the role of the government in providing services related to information, marketing facilities, technical assistance, and laws and regulations related to standards and quality control. Lastly, the WTO offers an opportunity for China and India to join hands and create a third bloc of countries besides the European Union and the United States in trade negotiations.

Improve Health, Education, Infrastructure, and Land Use at an Early Stage

The initial conditions of health, education, and land use also made a difference in the performance of reforms in China and India. In 1970 life expectancy was 49 years in India and 62 years in China; illiteracy affected nearly 70 percent of the Indian rural population compared with 49 percent in rural China. These differences may be accounted for by the fact that under the collective system in China, the government provided free basic health care and education to the rural population. After the start of reforms, both countries recorded a slowdown in the advancement of health and education. In India this was primarily due to the fiscal discipline imposed by the macroeconomic crisis, whereas in China market-oriented reforms introduced the logic of profit into the management of social services. This implied progressive privatization of supply agencies, a decline in government subsidies, and an increase in education and health costs, leading to an increase in school dropouts and in the health vulnerability of the population. In devising mechanisms to address the risks involved in the increased privatization of social services, China could perhaps learn from India's long experience with a vast array of government safety nets and welfare programs targeting the rural population.

China had also made more progress on rural infrastructure than India. Chinese government investment in power grew at 27 percent a year from 1953 to 1978, and rural electricity consumption grew at a rate of 27 percent a year from 1953 to 1980, then slowed to 10 percent a year from 1980 to 1990. In India rural infrastructure did not receive as much attention, particularly in the rural power sector, and thus rural electrification and the establishment of telecommunications connections proceeded more slowly in Indian villages. This slow pace severely affected the growth of agroprocessing and cold storage in the rural nonfarm sector. It is no wonder, therefore, that the levels of processing in Indian agriculture remain abysmally low.

In China the egalitarian access to land ensured by the land distribution and tenure system performed a crucial welfare function, providing the bulk of the rural population with access to a basic means of subsistence and limiting the number of landless. In India, on the other hand, land reforms to make the agrarian structure more equitable after independence were not as successful and left a relatively large number of landless agricultural labourers exposed to the negative consequences of unemployment and underemployment. Replicating the Chinese agrarian system does not seem politically feasible in India at this stage of development, so marginal and landless farmers will require a strong social protection system involving well-targeted social security and employment policies. Effective social protection measures will also be required in China, where land distribution is likely to become more skewed following the adoption of the new agricultural lease law that enables farmers to transfer lease rights and thus allows for the possibility of a higher concentration of land.

FURTHER REFORMS ARE NEEDED IN BOTH COUNTRIES

While both countries have made remarkable progress in terms of growth and poverty reduction, much remains to be done given the sizeable share of the population still living in poverty. The two countries are confronted with the formidable challenges of accelerating growth, improving efficiency, and ensuring that growth is equitable and sustainable.

Focus on Public Investments That Can Boost Agricultural Productivity Efficiently

Given the key role of agriculture in poverty reduction and

growth in China, public investments that boost agricultural productivity appear warranted. Significant increases in public investments seem unlikely because of budget pressures, so China and India will need to invest existing resources more efficiently. Studies have found that investments in agricultural research, education, and rural roads hold the greatest potential to promote agricultural growth and poverty reduction in both countries.

Farmers will have little potential to increase the amount of land they cultivate, so agricultural research and technology development is needed to help them increase agricultural growth by boosting their yields. Agricultural R&D takes place in both the public and the private sectors, but managing public versus private agricultural R&D can be tricky. In a bid to increase research funding, China promoted the development of the public business sector through commercialization of technologies by public research institutes. This approach often led, however, to the duplication of research with state-owned traditional research institutes. Improved intellectual property rights (IPR) regimes have stimulated private research and patenting activity in both countries. However, weak implementation of IPR in both countries and the high costs of maintaining patents in China are obstacles to the entry of new private players.

Significant opportunities for public-private partnerships can arise in the areas of funding, improving efficiency, and extension. The private sector, however, tends to favour higher-value crops and concentrate in areas where agriculture is already advanced. Given the potential of agricultural research for poverty reduction in marginal regions, public research spending should focus on addressing the needs of poorer farmers in less-favoured environments, such as India's semiarid tropics and rainfed areas and China's poor western regions.

Past government spending on irrigation, dominated by creation of large surface irrigation schemes, played an important role in promoting agricultural growth and poverty reduction, but today similar spending has smaller marginal returns, in terms of both growth and poverty reduction. It might be the case that investment in rainfed areas or traditionally lower-potential areas has higher returns today. Indeed, studies have shown that investments in rainfed areas of both countries have had high marginal returns for agricultural growth and poverty reduction. So major investments in harvesting rainwater through watersheds, through public-private partnerships, may help usher in a "multicoloured revolution" (not just a "green" one) in agriculture. In both countries there is also vast scope for improving water use efficiency through institutional and management reforms of the existing water systems. India has had useful experiences with water user associations in some selected states, participatory watershed schemes, and community-based rain harvesting. But these successful experiments need to be scaled up to make a significant difference for agriculture growth and poverty reduction. In China providing irrigation system managers with incentives to improve user efficiency had a positive effect on crop yields, the groundwater table, and cereal production.

Providing the right incentives to farmers is crucial to promote water saving. Low water prices and profligate subsidies on power for operating tubewells encouraged wasteful use of water and depletion of groundwater resources. Ambiguous water use rights following decollectivization in China, and laws linking water rights to land ownership in India, also led to inefficiencies. For example, unfair water markets emerged over time, in which rich landholders who can afford modern water extraction technology profit by selling water to poorer cultivators. Increases in water use charges may not be feasible in the short to medium term, however, without changes in the institutional environment.

Another distinctive pattern among the two countries in the past two decades is the much higher savings rates in China (about 45 to 50 percent) than in India (about 25 to 30 percent). The high Chinese savings rates, which facilitated boosting investments, are a puzzle in international comparisons. They might have been stimulated by high expected returns, including from investments in education, a matter which warrants further research.

Promote Rural Diversification and Vertical Coordination

A major shift in farm production toward non-foodgrain products such as livestock, fish, and horticulture has been well under way in India and China since the 1980s. The experience of China shows that achievement of food self-sufficiency and the extraordinary growth in basic grain production experienced by the late 1970s was a necessary precondition for diversification. The availability of food surpluses provided the government with enough leeway to feed the increasing population and relax controls over the foodgrain sector. Once food self-sufficiency was achieved, China gradually abandoned the policies biased in favour of rice and wheat, encouraging farmers to diversify production. In India, on the other hand, rising minimum support prices artificially boosted production of major cereals, discouraging diversification of production toward nongrain commodities. Moreover, policymakers must step up investment in research and infrastructure for high-value products such as livestock and horticulture to boost yields and expand their cultivation and processing, given their export potential, positive impact on smallholders, and growing domestic demand.

Rising consumer demand for non-foodgrain products is a major force driving diversification. Without vertical coordination of production, processing, and marketing--that is, between "plow and plate"--the potential for growth inherent in the diversification process is likely to remain underexploited. Both countries must strengthen the innovative institutional arrangements that have emerged to promote the development of new products. India's successful experience with contract farming in reducing risks, promoting the production and export of high-value foods, and increasing the income and employment of smallholders could be valuable for China. China's experience with growth in retail food chains and supermarkets in recent years could benefit India, where restrictions on foreign investment and infrastructure bottlenecks are limiting development in these areas.

Another dimension of rural diversification is provided by the evolution of a vibrant rural nonfarm sector. China's experience is instructive. The rapid growth of rural enterprises in China was a critical factor in the success of its reforms, especially in relation to poverty reduction. China's township and village enterprises (TVEs) provided increasing job opportunities outside agriculture, thereby diversifying and expanding the sources of household income. TVEs benefited from the close connection with urban markets that had been established since the early stages of their development.

India's nonfarm economy primarily produces for the rural population and markets and is dominated by tiny, family-operated units. These firms have low productivity because of a poor technological base and policies aimed to protect rural employment by reserving certain activities for small-scale units. Limited growth of rural nonfarm job opportunities in India is also related to the lack of knowledge and skills on the part of the poorly educated rural labour force.

The role of nonfarm employment is expected to become increasingly significant in the context of smallholder agriculture as the average farm size gets smaller. Greater off-farm opportunities and migration to urban areas are required to increase average farm size as well as labour productivity and farmers' income.

Use Well-Targeted Anti-poverty Programs and Safety Nets to Help the Poorest

The role of safety nets in poverty alleviation came into focus during the 1990s as China and India recognized the need to address the negative effects of liberalization policies on income distribution. Poverty funds and programs have documented shortfalls and inefficiencies in terms of targeting and cost-effectiveness, but they have contributed significantly to limiting the severity and the extent of poverty. There are still more than 300 million rural poor in India and China, based on the international standard of one dollar a day (more than 100 million in China and more than 200 million in India).

Antipoverty programs can be more practical and agile instruments for tackling poverty in the short run than public investments or radical redistributive measures such as land reforms. Given the fiscal discipline imposed by macroeconomic stabilization reforms, however, it is crucial to address the shortcomings of antipoverty programs. The experience of India shows that using a variety of targeted programs directed to specific sections of the poor can help improve targeting compared with the broader income- or area-based approaches traditionally implemented in China.

Decentralized and participatory approaches are more effective at strengthening the impact of antipoverty programs than top-down strategies and involve a greater variety of agents (NGOs, civil society, and international aid) in the fight against poverty besides the government. In India the extensive participation of panchayats (forms of local government with heavy public participation) and civil society at various stages of formulating and implementing antipoverty programs ensures that programs are tailored to local needs and can be carried out without extensive leakage.

Work to Make Governance Both Effective and Transparent

In both countries there was political will to carry out reforms, but in practice, outcomes have been shaped by the different patterns of governance. India is a "debating society" in which political differences are expressed freely, policymaking is exposed to pressure by various interest groups, and there are thus long debates before decisions are made. Subsequently, implementation is slowed by the lengthy bureaucratic procedures, set up to ensure checks and balances. This exercise, while compatible with the needs of a free and dynamic polity, considerably slows the pace of economic reforms. China, in contrast, is a "mobilizing society" in which decisions are made faster and state power is backed by mass mobilization. As a result, implementation of decisions is more effective, although the lack of extensive debate in China on major changes and reforms can also lead to disastrous courses of action, such as the "Great Leap Forward" in 1958, which resulted in massive famine, and the Cultural Revolution from 1966 to 1976. As the economic system opens up further and prosperity increases, it will become harder and harder to reconcile the centralized political setup with the more liberal economic system, and this is one of China's most important challenges today.

Although investments in rural infrastructure and other key public services are crucial, it is equally critical to develop suitable institutional arrangements for their delivery. In both countries the government is the major supplier of infrastructure services, but there are major failures and inefficiencies in provision owing to the lack of transparency and accountability. Strengthening the public institutions that provide public goods and services can lead to both fiscal sustainability (through significant cost reductions) and long-term growth (through improved quality of services provided). These goals can be achieved in different ways, including privatization, unbundling, decentralization, and contracting. Effective public institutions also require an adequate supply of trained and motivated personnel, as well as investments in training to help increase the supply.

Reforms have also been slowed at the implementation level by the regulatory environment and enforcement bureaucracy. In India, many inefficiencies remain in place, although reforms, including de-licensing, have been introduced to streamline the regulatory apparatus. During the reform years China relaxed regulations on mobility between rural and urban areas, which gave impetus to the development of the nonfarm sector and increased migration for economic purposes. In recent years the Chinese government has also started to relax the complex system of regulations affecting broad-based personal mobility.

Finally, with regard to the political systems, effective implementation of reforms in China was facilitated by a high level of centralization of decisionmaking, which minimized dissent. In the context of a democratic system and highly pluralist society such as India, consent is more difficult to achieve, and it is much more difficult to set clear objectives or timeframes for transition (such as for phasing out subsidies, reducing tariffs, or increasing prices). This situation slows the pace of change in the short and medium run. Although democracy and participation have intrinsic value and are not just instruments of development, the role of democracy in enhancing or hampering economic change and poverty reduction remains a complex subject for development research. Comparisons of China and India on these broad political matters may produce a fascinating set of insights in the coming years.

CONCLUSION

A number of factors help to explain the difference in growth during the pre-reform era: initial conditions, the sequencing and pace of reforms, and the political system, institutions, and regulatory environment. Yet special mention must be made of the fact that China and India achieved remarkable development and growth even as aid as a percentage of GDP in the two countries remained low. This is in direct contrast to most other developing countries and regions, where aid is much higher but commensurate development and poverty reduction outcomes have not been realized. This fact bears an important lesson for developing and developed countries, multilateral agencies, and local NGOs and groups. It questions the very basis of current policy prescriptions that accompany aid packages, not only raising issues related to the efficiency and effectiveness of external aid but also, conversely, revealing the extraordinary and often underestimated capacity of national initiatives and policy actions to halt--and in fact turn--the tide of poverty.

Both countries now face tremendous challenges on the path to further prosperity. Continued growth is a must, owing to pressure from population growth and the need for employment. It is also a condition for a more stable society. Given the high expectations of their citizens, the lack of growth or even slower growth could lead to unrest in both countries. The limited natural resource base can be a critical constraint to growth. The future economic growth of both countries increasingly depends on imports of energy, for which future prospects are uncertain. Both countries are also among those most severely affected by water shortages. Consequently, future growth must be based on higher efficiency and will require China and India to invest in science and new technologies to harness energy and water, optimize their economic structures for allocative efficiency, and reform their fiscal, financial, banking, and insurance systems. Both countries must also pursue more pro-poor growth, which is not only a development objective in itself, but also a precondition for future growth in the long term.

China and India can both gain tremendously by learning from each other, as both nations still face a long road ahead. The dragon has attained height and the elephant is starting to gather momentum, but both need to address their weaknesses and build on their strengths in order to achieve their national goals and fulfil the aspirations of their people. The lessons learned from the experiences of China and India are also of relevance to other developing countries and the fight against global hunger and poverty.

AGRICULTURE EXTENSION SYSTEM OF INDIA

After independence in 1947, the government's first step toward building an agricultural extension system was expansion of the World War II Grow More Food Campaign. Administrators and extension workers were exhorted to convince cultivators of the gains in yields that could be obtained through the use of improved seeds, compost, farmyard manure, and better cultivation practices. Rural agents, often inundated with other assignments, had little or no training for extension work, however, gains in yields were minimal, and India's leaders came to realize that converting millions of poor farmers to the use of new technologies was a colossal task.

The Community Development Programme in India was inaugurated in 1952 to implement a systematic, integrated approach to rural development. The nation was divided into development blocks, each consisting of about 100 villages having populations of 60,000 to 70,000 people. By 1962 the entire country was covered by more than 5,000 such blocks. The key person in the program was the village-level worker, who was responsible for transmitting to about ten villages not only farming technology, but also village uplift programs such as cooperation, adult literacy, health, and sanitation. Although each block was staffed with extension workers, the villagers themselves were expected to provide the initiative and much of the needed financial and labour resources, which they were not in a position to do or inclined to do. Although progress had been made by the early 1960s, it was apparent that the program was spread too thin to bring about the hoped-for increase in agricultural production. Criticism of the program led to more specialized development projects, and some of the functions were taken up by local village bodies. There was only a negligible allocation for community development in the sixth plan, however, and the program was phased out in the early 1980s.

The Intensive Agricultural District Programme, launched in five districts in 1960 by the central government in cooperation with the United States-based Ford Foundation, used a distinctly different approach to boosting farm yields. The program operated under the premise that concentrating scarce inputs in the potentially most productive districts would increase farm-crop yield faster than would a wider but less concentrated distribution of resources in less productive districts. Among these inputs were technical staff, fertilizers, improved seeds, and credit. Under the technical guidance of American cooperative specialists, the program placed unusual emphasis on organizational structures and administrative arrangements. For the first time, modern technology was systematically introduced to Indian farmers. Within a decade, the program covered fifteen districts, 28,000 villages, and 1 million inhabitants. The Intensive Agricultural District Programme was thus a significant influence on the forthcoming Green Revolution.

IRRIGATION IN INDIA

Except in southeastern India, which receives most of its rain from the northeast monsoon in October and November, dryland cultivators place their hopes for a harvest on the southwest monsoon, which usually reaches India in early June and by mid-July has extended to the entire country. There are great variations in the average amount of rainfall received by the various regions--from too much for most crops in the eastern Himalayas to never enough in Rajasthan. Season-to-season variations in rainfall are also great. The consequence is bumper harvests in some seasons, crop-searing drought in others. Therefore, the importance of irrigation in India cannot be overemphasized.

Irrigation in India has been a high priority in economic development since 1951; more than 50 percent of all public expenditures on agriculture have been spent on irrigation alone. The land area under irrigation expanded from 22.6 million hectares in FY 1950 to 59 million hectares in FY 1990, an increase of 161 percent in four decades. This increase was about 33 percent of the estimated potential. The overall strategy has been to concentrate public investments in surface systems, such as large dams, long canals, and other large-scale works requiring huge outlays of capital over a period of years, and in deep-well projects that also involve large capital outlays. Shallow-well schemes and small surface-water projects, mainly ponds (called tanks in India), have been supported by government credit but were otherwise installed and operated by private entrepreneurs. Roughly 42 percent of the net irrigated area in FY 1990 was from surface water sources. Tanks, step wells, and tube wells provided another 51 percent; the rest came from other sources.

Between 1951 and 1990, nearly 1,350 large- and medium-sized irrigation works were started, and about 850 were completed. The most ambitious of these projects was the Indira Gandhi Canal, with an anticipated completion date of close to 1999. When completed, the Indira Gandhi Canal is the world's longest irrigation canal. Beginning at the Hairke Barrage, a few kilometers below the confluence of the Sutlej and Beas rivers in western Punjab, it is running south-southwest for 650 kilometers, terminating deep in Rajasthan near Jaisalmer, close to the border with Pakistan. A dramatic change already had taken place in this hot and inhospitable wasteland by the late 1980s. As a result, desert dwellers switched from raising goats and sheep to raising wheat, and outsiders flocked in to purchase six-hectare plots for the equivalent of US$3,000.

Progress in irrigation has not been without problems. In India, arge dams and long canals are costly and also highly visible indicators of progress; the political pressure to launch such projects was frequently irresistible. But because funds and technical expertise were in short supply, many projects moved forward at a slow pace. The Indira Gandhi Canal project is a leading example. And the central government's transfer of huge amounts of water from Punjab to Haryana and Rajasthan, frequently cited as a source of grievance by Sikhs in Punjab, contributed to the civil unrest in Punjab during the 1980s and early 1990s.

Problems also have arisen as ground water supplies used for irrigation face depletion. Drawing water off from one area to irrigate another often leads to increased salinity in the supply area with resultant effects on crop production there. Some areas receiving water through irrigation are poorly managed or inadequately designed; the result often is too much water and water-logged fields incapable of production. To alleviate this problem, more emphasis is being placed on using irrigation water to spray fields rather than allowing it to flow through ditches. Furthermore, charges of corruption and mismanagement have been levied against government-operated facilities. Cases of bribery, maldistribution of water, and carelessness are frequently raised in the media.

Another major problem has been the displacement of thousands of people, usually poor people, by large hydroelectric projects. Critics also claim that the projects are damaging to the ecology. Smaller projects and such traditional methods for irrigation as tanks and wells are seen as having less serious impact. In the late 1980s and early 1990s, the debate between large-scale versus small-scale projects came to the fore because of the US$3 billion Sardar Sarovar project on the Narmada River. Sardar Sarovar, as conceived, was one of the world's largest hydroelectric and irrigation projects. Some 37,000 hectares of land in Madhya Pradesh, Gujarat, and Maharashtra were slated to be submerged following the construction of some 3,000 dams, 75,000 kilometers of canals, and an electric power generating capacity of 1,450 megawatts of power per year. Included among the 3,000 dams was the proposed 160-meter-high Sardar Sarovar Dam. In 1985 the World Bank agreed to loan US$450 million for the project. Environmentalists in India and abroad, however, argued that the project was ecologically undesirable. In the face of this strong protest, the World Bank appointed a two-member team in 1991 to review the project. Despite a negative review of the environmental impact by the team, World Bank funding and the project continued. By 1993, however, in the face of continued international protest as well as opposition and a call for a satyagraha by villages in the affected areas, the central government cancelled the dam project loan. Work on the Sardar Sarovar project continues, however, with funds provided by the central government and the governments of the three states involved.

Although India had the second largest irrigated area in the world, the area under assured irrigation or with at least minimal drainage is inadequate. The irrigation potential estimated to have been created by the early 1990s was about 82.8 million hectares. This amount includes the gross irrigated area plus the potential for double cropping provided by irrigation. There was a cumulative gap in irrigated land use of about 8.6 million hectares until FY 1990, by which time the gap had decreased through improved land management.

3

Impact of Economic Reforms on Indian Agricultural Sector

INTRODUCTION

The existence or absence of favourable natural resources can facilitate or retard the process of economic development. Professor W.A. Lewis writes : "Natural resources determine the course of development and constitute the challenge which may not be accepted by the human mind". Developing countries, embarking on programmes of economic development, "usually have to begin with and concentrate on the development of locally available natural resources as an initial condition for lifting local levels of living and purchasing power, for obtaining foreign exchange with which to purchase capital equipment, and for setting in motion the development process". With the basic thrust on higher growth in food grain production and other agricultural commodities, increase in productivity and efficient use of resources in agriculture has received special emphasize all through the process of the development, since independence. Sustainable agricultural production depends on the judicious use of natural resources (soil, water, livestock, plant genetic, fisheries, forest, climate, rainfall, and topography) in an acceptable technology management under the prevailing socio-economic infrastructure. Food and Agriculture Organization (FAO) has formulated the following definition for sustainable development in the context of agriculture, forestry and fisheries:

> "Sustainable development is the management and conservation of the natural resource base and the orientation of technological and institutional change in such a manner as to ensure the attainment and continued satisfaction of human needs for the present and future generations. Such sustainable development (in the agriculture, forestry and fisheries sectors) conserves land, water, plant and animal genetic resources, is environmentally non-degrading, technically appropriate, economically viable and socially acceptable".

ECONOMIC REFORMS PROCESS

Since July, 1991 the country has taken a series of measures to structure the economy and improve the balance of payments position. The New Economic Policy (NEP-1991) introduced changes in the areas of trade policies, monetary and financial policies, fiscal and budgetary policies, and pricing and institutional reforms. The salient features of NEP-1991 are:

(i) liberalization (internal and external),
(ii) extending privatization,
(iii) redirecting scarce Public Sector Resources to Areas where the private sector is unlikely to enter,
(iv) globalization of economy, and
(v) market friendly state.

Research reports reveal that this macro-economic adjustment programme is remarkable for its relatively painless transition compared with similar programmes elsewhere and a large part of the credit for absorption of these shocks is due to the steady increase in agricultural production. The GATT Agreement signed in 1995 will fundamentally change the global trade picture in agricultural sector.

IMPACT OF ECONOMIC REFORMS PROCESS ON INDIAN AGRICULTURAL SECTOR

Agricultural sector is the mainstay of the rural Indian economy around which socio-economic privileges and deprivations revolve, and any change in its structure is likely to have a corresponding impact on the existing pattern of social equality. No strategy of economic reform can succeed without sustained and broad based agricultural development, which is critical for :

(i) raising living standards,
(ii) alleviating poverty,
(iii) assuring food security,
(iv) generating buoyant market for expansion of industry and services, and
(v) making substantial contribution to the national economic growth.
(vi) Studies also show that the economic liberalization and reforms process have impacted on agricultural and rural sectors very much.

According to Bhalla, of the three sectors of economy in India, the tertiary sector has diversified the fastest, the secondary sector the second fastest, while the primary sector, taken as whole, has scarcely diversified at all. Since agriculture continues to be a tradable sector, this economic liberalization and reform policy has far reaching effects on:

(i) agricultural exports and imports,
(ii) investment in new technologies and on rural infrastructure
(iii) patterns of agricultural growth,
(iv) agriculture income and employment,
(v) agricultural prices and
(vi) food security.

Reduction in Commercial Bank credit to agriculture, in lieu of this reforms process and recommendations of Khusrao Committee and Narasingham Committee, might lead to a fall in farm investment and impaired agricultural growth. Infrastructure development requires public expenditure which is getting affected due to the new policies of fiscal compression.

Liberalization of agriculture and open market operations will enhance competition in "resource use" and "marketing of agricultural production", which will force the small and marginal farmers (who constitute 76.3% of total farmers) to resort to "distress sale" and seek for off-farm employment for supplementing income.

MARGINALISATION OF SMALL FARMERS

A central issue in Agricultural Development is the necessity to increase productivity, employment, and income of poor segments of the agricultural population. Among the rural poor, the small farmers constitute a sizeable portion in the developing countries. Studies by FAO have shown that small farms constitute between 60-70% of total farms in developing countries and contribute around 30-35% to total agricultural output.

Liberalisation era (1990-91) began in India when over 40% of rural households were landless or near landless, and over 96% of the owned holdings and 68.53% (over 2/3rd) of owned land belonged to the size groups (marginal, small and semi-medium). The decade of 1981-82 to 1991-92 seems to have witnessed a marked intensification of the marginalisation process - the percentage of small owners increased from 14.70% to 21.75%.

Small farmers emerged as the size group with the largest share of 33.97% in the total land, which is just doubled during this decade. As regards the large farmers, they were 1 % of the total owners in 1990-91 but owned nearly 13.83% of the total land. An interesting, but speculative, inference is that the changing position of the large owners represents the other side of the marginalisation process, i.e., the presence, and possibly growing strength, of a small but dominant and influential group in agriculture. Analytical reports reveal that marginalisation process could gather further momentum in the years ahead to become an explosive source of economic and political turbulence, due to the features of prevailing policy-cum-market environment in the country.

Trend towards a greater casualisation (erratic and low-paid work) of the workforce that was witnessed in the 1980s appears to have continued in the1990s. Low productivity and inability to absorb the growing labour force make the agricultural sector in India witness to a pervasive process of marginalisation of rural people. This process is likely to get intensified in the coming years, raising formidable problems in achieving sustained development of rural areas and rural people.

Both Information Technology, Genetic Engineering and Bio-Technology, which are the "drivers" of globalisation with their complementarities of liberalisation, privatisation and tighter Intellectual Properties Rights, are bound to create new risks of marginalisation and vulnerability. Information Technology is able to produce a penetrating and clinical mapping of the land, encompassing the physical, chemical and biological features, and groundwater resources, and forecast of climatic conditions in a focused manner, that even small geographical segments—the small farms—can be benefited through the guidance provided by the ways in which natural and human resources can be optimally combined with appropriate technologies, inputs and options to enhance and diversify agricultural production. Information Technology will facilitate dissemination of information on development, education, extension, husbandry, marketing, production, and research, to agricultural farmers.

INDIAN AGRICULTURAL SECTOR

The Indian Agricultural sector provides employment to about 65% of the labour force, accounts for 27% of GDP, contributes 21% of total exports, and raw materials to several industries. The Livestock sector contributes an estimated 8.4% to the country GDP and 35.85 % of the agricultural output. India is the seventh largest producer of fish in the world and ranks second in the production of inland fish. Fish production has increased from 0.75 million tons in 1950-51 to 5.14 million tons in 1996-97, a cumulative growth rate of 4.2% per annum, which has been the fastest of any item in the food sector, except potatoes, eggs and poultry meat. The future growth in agriculture must come from viz.,

(i) new technologies which are not only "cost effective" but also "in conformity" with natural climatic regime of the country;

(ii) technologies relevant to rain-fed areas specifically;

(iii) continued genetic improvements for better seeds and yields;
(iv) data improvements for better research, better results, and sustainable planning;
(v) bridging the gap between knowledge and practice; and
(vi) judicious land use resource surveys, efficient management practices and sustainable use of natural resources.

Agricultural Planning and Development

India is a vast country with a variety of landforms, climate, geology, physiography, and vegetation India is endowed with regional diversities for its uneven "economic and agricultural" development, on account of :

(i) Agro-climatic environments (15 Zones/127 regions),
(ii) Agro-ecological regions (20) and 60 sub-regions,
(iii) Agro-Edephic regions,
(iv) Terrain mapping sub-units,
(v) Natural resources endowments (geology, geomorphology, soil, ground water, surface water, & infrastructure),
(vi) Human resources (Population density),
(vii) Level of investments in rural infrastructure, and
(viii) Level of investment in technology and its adoption.

India has a total geographical area (TGA) of 329 Million Hectares (MH) out of which, about 265 MH represent varying degrees of potential for biological production. A report reveals that more than 50% of TGA is threatened by various types of land degradation, such as soil erosion, gully and ravine formation, salinity, water logging, shifting cultivation, etc. Development of irrigation potential is considered as the key factor in the sustenance of "Green Revolution". Despite 50 years of development planning, rainfed agriculture is the largest and the most important sector of crop production in India.

Soil resources are the most precious non-renewable vital resources for growing food, fibre, and fuel wood to meet the human needs. Management of Soil Resources is essential for both the continued agricultural productivity and protection of environment. By considering various factors like population growth rate, diminishing per capita of land and water resources, and increasing land degradation problems, it is estimated that India will be required to produce an additional 5-6 million tons of food grains annually in 21st Century. This will lead to tremendous pressure on soil resources along with competitive demand for it from industrialization and urbanization. However the capacity of soil to produce is limited and its limits to production are set by its inherent characteristics, agro-ecological settings, and its use and management.

Forests are an important natural resources of India, having a moderating influence against floods and also protecting the soil against erosion. About 95% of the forests in India are owned by States and the total area under forests is about 22% of the total geographical area.

Development of livestock has been envisaged as an integral part of sound system of diversified agriculture. In animal production, the major aim is for raising ecologically adapted animals and efficient utilization of locally available feed resource. Dairy development is intimately linked with cattle population, breed improvement, cattle health and disease management, and fodder development, etc. Animal Husbandry in India is essentially a endeavour of millions of small holders (Resource-Poor-Farmers) who rear animals on "crop residues" and "common property resources" without generally allowing them to compete with man for food grains.

The small holders produces milk, meat, wool, etc., for the community, with virtually no capital, resource, training and at a cost that no modern technology in the world had ever produced. Food and Fodder Resources will be crucial to the future development of "livestock resources" in the Country. There is very little scope for increasing the area under fodder production, keeping in view the priority for food grains, pulses and oil seeds. Development of Fodder Resources is basically an activity based on a multi-disciplinary approach involving the areas of agriculture, animal husbandry, environment & forests, revenue, rural development, and wasteland development.

Water Resources of India contain diverse group of flora and fauna. Agriculture is the greatest user of Water accounting for about 80% of all consumption. Animal Husbandry and Fisheries require abundant water. Development of Water Resources, since Independence, has been undertaken for specific purposes like irrigation, flood control, hydro-power generation, drinking water supply, industrial and various miscellaneous uses. Minor irrigation projects have both surface and ground water as their source, while major and medium projects mostly exploit surface water resources. The break up of the ultimate irrigation potential under the three categories are :

58 M.Ha by major and medium irrigation projects,
17 M.Ha by minor surface water schemes, and
64 M.Ha by minor ground water schemes.

Fisheries Resources of India are either inland or marine. The principal rivers and the tributaries, canals, ponds, lakes, reservoirs comprise inland fisheries. The river extend about 27,200 kms, and other subsidiary water channel comprise about 112,000 kms. Marine resources comprises of about 2 Million sq.kms of EEZ for deep sea fishing, and 7,250 kms of coastline. With the diverse fish fauna, the development objectives are to judiciously and optimally utilize the resources for [NBFGR2K]:-

(i) enhancing production and productivity of fishermen, fish farmers and fishing industry;
(ii) increasing fish production and thereby, raising nutritional standard of people;
(iii) earning of foreign exchange from export of marine products;
(iv) improving Socio-economic conditions of traditional fishermen;
(v) generating employment for coastal and rural poor; and
(vi) conservation of depleting species of fish.

Good infrastructure helps in raising productivity and lowering the unit cost in the production activities of the economy. "Agricultural Infrastructure" refers to "Rural Infrastructure" whereas "Industrial Infrastructure" refers to "Urban Infrastructure". Agricultural development requires:

(i) agricultural research and extension,
(ii) rural financial institution,
(iii) irrigation and drainage,

(iv) agricultural inputs (fertilizers, seeds, credits), and
(v) marketing and storage facilities.

Agriculture Credit is a crucial input for increasing agricultural production and productivity. Institutional finance for Agricultural credit is disbursed mainly by Commercial banks, Regional Rural Banks, Land Development Banks, and Cooperative banks. Share of commercial banks in total institutional credit to agriculture is about 48%, that of Cooperative banks is about 46%, and Regional Rural Banks account for 6% only. Short-term Credit accounts for 2/3rd of the total institutional lending to the Agriculture.

Drought has multiplier effect on agricultural production during the subsequent year also, due to:

(i) non-availability of quality seeds for sowing of crops,
(ii) inadequate draught power for carrying out agricultural operations as a result of either distress sale of cattle or loss of life,
(iii) reduced use of fertilizers as the investment capacity of the farmers decline,
(iv) non-availability of raw materials in agro-based industries, and
(v) deforestation to meet the energy needs in domestic sector as agricultural waste may not be available in required quantity.

The Central Ministry of Agriculture (MOA) is responsible for implementation and formulation of national policies and programs to achieve agricultural growth through optimum utilization of the land resources, water, soil, plant, fisheries, and livestock resources. Government of India implements the following agricultural related Schemes (whether Watershed based or Agro-climatic region based) in the country, which deal agricultural resources information for Planning and Development:

(i) Agro-climatic Regional Planning (ACRP) Project
(ii) Agro-Ecological Mapping Project of the National Bureau of Soil Survey and Land Use Planning (NBSS&LUP)
(iii) All India Soil and Land Use Survey (AISLUS)

(iv) Early Warning System of Agricultural Situation in India

(v) Forecasting of Agricultural output using Space, Agro-meteorology and Land based observations (FASAL) Project

(vi) Land Records Computerisation Project

(vii) National Agricultural Research Project (NARP)

(viii) National Agricultural Technology Project (NATP) to strengthen research-extension-farmer (r-e-f) linkage

(ix) National Watershed Development Program for Rain-fed Areas (NWDPRA)

(x) Soil and Water Conservation Programs

(xi) Drought Prone Area Development programme

(xii) Desert Development Programme

(xiii) National Wastelands Development programme

(xiv) Integrated Mission on Sustainable Development (IMSD) Programme

(xv) Information for decision making

The major objective of Sustainable Agriculture and Rural Development is to increase food production in a sustainable way and enhance food security. The Agenda-21 recommends major adjustments in agricultural, environmental and macro-economic policy to create the conditions for the Sustainable Agriculture and Rural Development. Recommendations of the United Nations Conference on Environment and Development—Agenda 21 (1992) on "Information for decision making" are as follows:

(i) development of indicators for sustainable development,

(ii) promotion of global use of indicators for sustainable development,

(iii) improvement of data collection and use,

(iv) improvement of methods of data assessment and analysis,

(v) establishment of comprehensive information framework,

(vi) strengthening of capacity for traditional information,

(vii) production of information usable for decision making,

(viii) development of documentation about information,
(ix) establishment of standards and methods for handling information,
(x) establishment and strengthening of electronics networking capabilities, and
(xi) making use of commercial information sources.

An Informatics model will have the knowledge components such as objects, events, know-how, precedence and cause-and-effect relationships and Meta-knowledge. Informatics, which is an IT application, is taking advantage of:

(i) multi databases (Federated and non-Federated databases),
(ii) information system research and development methodology,
(iii) relational-object methods,
(iv) knowledge base and expert systems,
(v) Geographical Information System (GIS) Technology,
(vi) model bases,
(vii) distributed query capabilities over INTERNET/INTRANET.

Development of Information Systems and utilization of Information Resources over INTERNET/INTRANET is a matter of strategic importance in all countries today. Informatics Network plays an important role in the information flow from the implementation level to the planner at Macro (national) level, Macro-meso (region covering more than one state) level to Meso (state) level, and Micro (District, Block and Village) level.

Metadata standards are simply a common set of terms and definitions that describe geospatial and non-spatial data. Metadata standards provide a way for data users to know:

(i) What data are available
(ii) Whether the data meet specific needs
(iii) Where to find the data
(iv) How to access the data

The information needed to create metadata is often readily available, when the data are collected. A small amount of time invested at the beginning of a project may save money in future. The initial expense of documenting data clearly outweighs the potential costs of duplicated or redundant data generation. Metadata organization will facilitate for internet access to distributed sites where data are produced, maintained or used.

The OpenGISÒ Model of the Open GIS Consortium Technical Committee [OpenGIS] envisages to synchronize geo-processing technology with the emerging Information Technology standards, based on open systems, distributed processing, and componentware frameworks, and to facilitate interoperability through "common specification" over internet/ intranet. The "Pluggable Computing Model" provides a conceptual framework ("reference model") that positions the OpenGIS Specification in the broad context of Information Technology. The Pluggable Tool Services include GIS Tools, Imaging Tools, Expert Tools, and RDBMS Tools. Each Tool has algorithms, data, and interfaces to services in the distributed computing environment. Benefits of the Pluggable Computing Model are as follows:

(i) To permit increased resource sharing between organizations and processes

(ii) To facilitate understanding the role of the OpenGIS Specification in the larger context of Information Technology

(iii) To enhance data connectivity among users and applications

(iv) To improve the ability of developers and users to integrate new capabilities into existing environments as well as incorporate legacy systems into new environments.

Informatics for agricultural development requires coordinated inter-sectoral approach and application of appropriate Information Technology (IT) tools, in the areas of :

(i) Agricultural Research,

(ii) Agro-meteorology,

(iii) Agricultural Marketing,

(iv) Agricultural Engineering and Food processing,

(v) Agricultural Extension and Transfer of Technology,

(vi) Credit & Co-operation,

(vii) Crop Production and Protection,

(viii) Environment and Forest,

(ix) Fertilizers and Manure,
(x) Fisheries,
(xi) Irrigation and Drainage Systems,
(xii) Livestock, Dairy Development and Animal Husbandry,
(xiii) Rural Development and Planning,
(xiv) Soil and Water Management,
(xv) Watershed Development, and
(xvi) Wastelands Development

In view of the recommendations given by ISDA-95 and various sub-Groups for formulation of the Ninth Plan in the Agriculture Sector, MOA is implementing Information Technology Plan, in collaboration with NIC, to implement "NICNET based Agricultural Informatics and Communication (AGRISNET)" in the country, to achieve higher sustainable agricultural productivity and also to make "Indian Agricultural Sector On-line". This is likely to be the largest sharable Internet Portal in the world, for agricultural sector in India, on NICNET having more than 10,000 nodes to government itself.

Agricultural Resources Information System

It is clear that sustainable agricultural production depends on the judicious mix of natural resources (soil, water, livestock, plant genetic, fisheries, forests, climate, rainfall, and topography) in an acceptable technology management under the prevailing socio-economic infrastructure. In addition to the natural resources components, it is also essential to combine natural resources with capital resources, institutional resources, and human resources for sustainable agricultural development. Agricultural Resources components include:

(i) Animal Resources
(ii) Capital resources
(iii) Climate resources
(iv) Environment data
(v) Fisheries Resources
(vi) Forestry Resources
(vii) Institutional resources
(viii) Land owners data
(ix) Plant Resources
(x) Socio-economic and Infrastructure data

(xi) Soil resources
(xii) Water Resources

For increasing production at micro level, an inventory of currently used, potentially available, and an evaluation of the quantity and quality of these resources is required. This requires design and development of agricultural resources information system using state-of-the-art IT Tools, as given below, to facilitate effective agricultural planning and development :

(i) Data warehousing (Data Bases & Model Bases)
(ii) Expert Systems & Knowledge Bases
(iii) Networking (Internet, Intranet and Extranet)
(iv) Geographical Information System (GIS)
(v) Application of Remote Sensing Data
(vi) Multi-media Information System
(vii) Decision Technology System
(viii) E-Commerce & E-Governance, and
(ix) Digital Library

Agricultural planning and development require (a) knowledge about recent progress in agriculture, (b) the existing situation (especially the main problems impeding development), and (c) the potentialities for achieving agricultural objectives. This information is needed for re-assessing current investment and other development activities as well as for planning new measures, setting benchmarks against which to monitor progress.

Proper analysis of the agricultural sector requires that it is seen as a system of functionality inter-related and inter-dependent elements, each of which contributes to the existing and potential level of performance of the sector. A stock taking and diagnostic survey is needed early in the planning process to provide information about the wide range of factors influencing agricultural performance.

Both the Ministry of Agriculture and Ministry of Rural Development implement, through corresponding State departments, various central sector and centrally sponsored schemes related to agricultural and rural development, on watershed basis. The landscape, climate, and agronomic characteristics of each watershed vary considerably. Each watershed contains a complex mixture of:

(i) soil types,
(ii) landscapes,
(iii) climatic regimes,
(iv) land use characteristics, and
(v) agricultural systems.

Each watershed can be subdivided into agro-eco-regions having similar soil types, landscapes, climatic regimes, crop and animal productivity, and hydrologic characteristics. Integrated Watershed Development and Management has been recognized as an effective strategy for sustainable agricultural development in the country.

Sources of Agricultural Resources Information and Design of System

Remote Sensing has provided a new impetus for the earth resource and environmental scientists. Increasing population and diminishing resources have compelled us to consider better ways for management of natural resources. Soil survey and preparation of soil maps are being carried out by NBSS&LUP, AISLUS, CAZRI, CSSRI, CSWCRTI, NRSA, RRSSC, IIRS, State Departments of Agriculture, State Soil Survey Units, State Agricultural Universities, State Remote Sensing Application Centres, etc.

A review of the soil mapping and land degradation mapping was conducted by an Inter-Agency Expert Committee constituted by the Ministry of Agriculture and the Department of Space, and on the basis of the recommendations, a National Mission on "Mapping of Soils and Land Degradation at 1:50,000 Scale" with the major objective of creation of uniform soil and land degradation database for the entire country is being contemplated. Forestry Survey of India, Geological Survey of India, Fisheries Survey of India, Botanical Survey of India, National Remote Sensing Agency, Survey of India, National Atlas and Thematic Mapping Organization, National Sample Survey Organization, Central Ground Water Board, etc., conducts resources surveys and develop "resources databases" using ground truths and applications of remote sensing data.

The Report of the Committee on "Natural Resources Information System (NRIS)—Linkage and Networking Project", constituted by the Department of Space in early 1990s, envisaged about 435 district level NRIS nodes in conjunction with DISNIC nodes of NIC, 26 state level NRIS nodes, 182 NRIS project nodes (7 Themes and 26 States), and 42 NRIS Regional nodes (7 themes and 6 regions). Development of "Natural Resources Information System (NRIS)—Linkage and Networking Project" was initiated by NIC in its pilot project districts. Department of Land Resources through its land resources development programmes, Department of Agriculture and Cooperation through its NWDPRA Projects, and Department of Science and Technology through its NRDMS Projects, have been involved in the implementation/ development of Natural Resources Information System (NRIS) to strengthen their schemes through their implementing agencies. The existing data available from the following reports can facilitate strengthening resources databases:

(i) Soil survey
(ii) Geological survey
(iii) Forest inventories
(iv) Hydro-meteorological studies
(v) Aerial photographs and contour maps
(vi) Ownership data and infrastructure information
(vii) Rainfall and stream flow data
(viii) Land use details
(ix) Development plans

Development of metadata is required as the overall rate of collection of data increases rapidly with advances in technologies such as high resolution satellite-borne imaging systems and global positioning system, and with growing number of people and organizations who are collecting and using data (spatial and non-spatial). Metadata standards on soil geographic data, vegetation geographic data, developed by [FCDC98], provide a systematic way to collect metadata.

Agricultural Resources Information System will have data and information on basic resources such as : (i) soil resources, (ii) water resources, (iii) climate resources, and other data sets (collated from Remote Sensing as well as conventional means) such as (iv) basic data on crops, (v) animal husbandry and fisheries, (vi) genetic (plant, animal and fisheries) materials, (vii) land ownership, (viii) Socio-economic data, (viii) infrastructure for agricultural development. The data sets are as follows:

(i) Basic Data on Crops
(ii) Production of major crops
(iii) Area cultivated under each major crop
(iv) Yields per Unit of Area for each crop
(v) Areas sown but not harvested
(vi) Areas of fallow, double cropped, irrigation and inter-cropped land
(vii) Information on livestock numbers, production and Yield per unit
(viii) Trade statistics on agricultural commodities and the extent to which imports/exports are involved
(ix) Information on size, character, technology and organization of farms, by groups
(x) The inventory and appraisal should cover natural, capital, institutional and human (manpower) resources.

Natural Resources

Information on physical feature [topography, geology, soils, natural vegetation, and hydrology (surface and sub-surface)] to determine the land's capability for agricultural development;

Maps depicting differences in physical land characteristics, meteorological, climatological, hydrological, geological, and geo-morphological conditions; population densities, types of land tenure systems used, proximity to markets and urban centres, transportation and other infrastructures;

Areas of immediate growth potential (where climate, soil and water conditions are favourable for agriculture and where technology needed to substantially increase output of major crops already being grown;

Areas of future growth potential (where favourable climatic and soil conditions exist but lack one or more elements of (i) adequate & controlled supply of water, (ii) technology required for substantially increasing production of a major crop or crops, currently grown, or capable of being grown, and (iii) transportation needed to bring the areas into national economy);

Areas of low growth potential (where climatological, soil, topological or other deficiencies without economic means for correcting them, exist) which require technological breakthroughs before substantial increases in output are possible.

Capital Resources

(i) Investments in agriculture (buildings, water systems, irrigation works, drainage systems
(ii) Agricultural implements and machinery
(iii) Work animals and breeding stock
(iv) Agricultural inputs (seeds, fertilizers, pesticides and insecticides, and credit)

Institutional Resources

(i) Research
(ii) Extension
(iii) Training
(iv) Provision of short, medium and long-term credits
(v) Marketing, and
(vi) Development plans

Human Resources (to find out what extent the human conditions act as a constraint on increased output and can contribute to increased output)

(i) Labour forces (owner-farmers, sharecroppers, and wage labourers)
(ii) Labour Force (employed, under-employed, and unemployed; seasonal variations)
(iii) Level of literacy, education, nutrition of agricultural population

Since sustainable agricultural development schemes are being implemented on watershed basis, database on watershed basis, as recommended by FAO, is required. AISLUS has brought out Watershed Atlas on 1:1,000,000 scale, which provides a uniform delineation and codification system of the watersheds that could be followed by all concerned agencies dealing with watershed approach, on a common basis. The whole country has been divided into six river resources regions, 35 basins, 112 catchments, 550 sub-catchments and 3,237 watersheds. FAO Reports on soil series recommend to have data on the following parameters with respect to each watershed:-

Database on Watershed

(1) General data
  (a) watershed name,
  (b) location,
  (c) boundaries,
  (d) size,
  (e) shape,
  (f) elevation,
  (g) slope,
  (h) the presence of streams, tributaries, etc.

(2) Physical data
  (a) geological data,
  (b) soil data (soil texture, depth, soil series, physical and chemical properties),
  (c) geomorphologic data (drainage patterns, stream density and order, channel profiles etc.)

(3) Climate, hydrology and water resource data
  (a) precipitation,
  (b) evaporation,
  (c) wind,
  (d) temperature,
  (e) humidity,
  (f) rainfall,
  (g) stream flow,
  (h) surface water,
  (i) ground water,
  (j) water quality, etc.

(4) Land Use, Land Cover, Land Capability data
  (a) land use by classes,
  (b) ownership,
  (c) land capability, etc.

(5) Cropping pattern, irrigation requirements, agricultural inputs availability, etc.

(6) Erosion data—
  (a) kinds of erosion, extent, causes, etc.

(7) Infrastructural data
  (a) transportation networks,

(b) housing, public services,
(c) survey of agro-industry, etc.

(8) Socio-economic/Demographic data
(a) population,
(b) rate of growth,
(c) composition,
(d) migration,
(e) employment, and
(f) other demographic factors affecting rate of resource use, etc.

(9) Minimum needs of the farmers
(a) more roads,
(b) domestic/irrigation water,
(c) housing,
(d) marketing arrangements,
(e) recreation facilities.

(10) Awareness of farmers about the causes and problems facing the watershed.

(11) Economic data
(a) farms production,
(b) farms income,
(c) farm models,
(d) farming systems,
(e) land use patterns,
(f) employment,
(g) labour demand and supply,
(h) rural enterprises,
(i) marketing, etc.

Maps for Developing DSS

(i) base map showing boundary, sub-watersheds, villages, roads, etc;
(ii) topographic map showing contours, elevations, land forms, streams, etc;
(iii) soil map showing soil types and boundaries, depths and soil limiting properties
(iv) climatic map showing mainly rainfall, but statistics may include temperature, evapo-transpiration, etc.;

(v) geology map showing rock types, structures, displacement, morphology, etc.;
(vi) slope map showing different slope classes or exposures/aspects;
(vii) present land use map showing major land uses and land cover types;
(viii) land capability or land suitability map showing different land capability classes; or land suitability classes;
(ix) land-use adjustment map showing land being over-used or under-used and adjustment needs;
(x) erosion or sediment source maps showing sites of various types of erosion and sediment potential areas;
(xi) hydro-meteorological network map showing the location of climatic and stream gauging stations;
(xii) water resource map showing surface and underground sources.

Decision Support Systems

Decision Support Systems (DSSs) can deliver the technology of management and the relevant knowledge which managers need to get their jobs done. Most of today's DSSs support the "choice" phase of decision making. Enhanced DSSs, known as Management Support Systems (MSSs), are attempting to fill the gaps in decision-making support, not provided by the traditional DSSs. For complete support of critical management activities of communication and decision making—including qualitative and creative processes—a collection of computer information system, as given below, is required:

(i) Decision Support System (DSS)
(ii) Group DSS (GDSS)
(iii) Executive Information System (EIS)
(iv) Executive Support System (ESS)
(v) Expert System (ES)
(vi) Idea Processing System (IPS)
(vii) Management Information System (MIS)
(viii) Office Automation System (OAS)
(ix) Transaction Processing System (TPS)

DSSs, Group DSSs, ESs, and components of ESSs and MISs primarily support the decision making process. "A complete set of enhanced DSS together with an IPS constitutes a Decision Technology System (DTS) which provides complete, integrated support for all phases of the decision-making process and delivers the complete technology of management for full decision support". Development of Decision Support Systems on the following areas facilitating Agricultural Resources Management are envisaged using the state-of-the-art IT tools :

(i) Crop Suitability based on factor endowment
(ii) Land Suitability Assessment;
(iii) Land Productivity Assessment;
(iv) Population Supporting Capacity;
(v) Land Evaluation and Land Use Planning;
(vi) Land Degradation Risk Assessment;
(vii) Quantification of Land Resources Constraints;
(viii) Land Management;
(ix) Agro-ecological Characterization for Research and Planning;
(x) Agricultural Technology Transfer;
(xi) Agricultural Inputs Recommendations;
(xii) Farming Systems Analysis and Development;
(xiii) Environmental Impact Assessment;
(xiv) Monitoring of Land Resources Development.
(xv) Livestock (cattle, buffalo, goat, & sheep) Farming Systems
(xvi) Water allocation in an irrigation system
(xvii) Fodder Resources Development
(xviii) Water Bodies (Basin) planning systems using Watershed and Agro-Eco Region Planning Concepts

There have been a lot of research publications and also software products developed by various Agricultural Research institutions, both in India and abroad. It has been envisaged to utilize these packages and develop DSSs in districts for agricultural resources management.

Application of Geomatics Technology

In order to reduce the risk of marginalisation and vulnerability of the small and marginal farmers, who constitute about 76.3% of total farmers of the country, it is suggested to develop "Agricultural Resources Information System", with the public funding, and make it available to the Resource-Poor-Farmer. Internet Technology based applications (Portals) on agricultural resources are expected to facilitate agricultural development, rural development, and backward area development in the country. Design of agricultural resources information system is to support the various management systems at National, State, District and Village level vis-a-vis catchment, watershed and micro watershed level.

Since MOA has formulated schemes to establish AGRISNET nodes in districts during the Ninth Plan Period and extend to block levels during the Tenth Plan Period, it is required to implement the Geomatics Technology Plan, as suggested below, in mission mode:

(i) Reaching-the-Unreached, "Resource-Poor-Farmers", through Information Technology applications;

(ii) "land information system" in district and block levels;

(iii) Strengthening "Agricultural Resources Information System" in all districts (regions) of the Country, "irrespective of past or future growth regions;

(iv) Development of decision support systems on "production practices and systems" which need to be adapted to respond to new market demands and export opportunities, poverty alleviation or growing labour shortages, depending on the agricultural production setting;

(v) Development of generic decision support systems (DSS) using databases and model bases for agricultural planning and management at micro watershed level; and also to establish GIS centres at block or panchayat level;

(vi) DSS on water allocation in an irrigation system to remove the existing disparities in the availability between the head-reach and tail-end farms and between large and small farms, to achieve "equity and social justice" ;

(vii) DSS on Land Resources development issues, given in the DSS section;

(viii) Linkages to Development of National Water Database as envisaged in the World Bank aided National

Hydrology Project to strengthen water resources management facilitating agricultural planning in districts;

(ix) Development of Agrometeorology Database providing vital information of long-term and short-term objective in agricultural production, planning and management;

(x) National Agricultural Drought Assessment and Monitoring System (NADAMS) and also Agromet Advisory Services on NICNET;

(xi) DSS on Water Bodies (Basin) based agricultural development, using Watershed and Agro-Eco Region Planning Concepts;

(xii) Development of "metadata" standards and application of "OpenGIS Model" on agricultural resources for Internet/Intranet access; and

(xiii) Involvement of Institutions viz., ICAR Institutions, State Agricultural Universities, Rural Development Institutes, NIC, DOS, NATMO, GSI, CGWB, Departments of Geography Research, etc., working on spatial data generation and application of spatial theory for problem solving in respect of agricultural development, rural development and backward area development

Conclusion

The central issue in agricultural development is the necessity to increase productivity, employment and income for poor segments of the agricultural population of whom the small and marginal farmers constitute a sizeable portion. Information Technology Tools viz., Data warehousing (Data Bases & Model Bases), Expert Systems & Knowledge Bases, Networking (Internet, Intranet and Extranet), Geographical Information System (GIS), Application of Remote Sensing Data, Decision Support Systems, and E-Commerce (b2b, b2c solutions), facilitate the Farmers to know the "agricultural situation" in Indian as well as abroad and accordingly undertake agricultural production.

This IT-led globalisation will certainly benefit the medium and large farmers who can invest on IT, as has happened during "green revolution". Since the agricultural development strategy has been mostly "growth-oriented" and therefore had a "built-in bias" in favour of Large farmers over Small farmers. Farmers can invest on computers to get access to Internet, but it is not possible for them to invest on "agricultural informatics" with decision support system using geomatics technology.

Agriculture being a "state subject" and a primary sector which accounts for about 27% of GDP, 65% of labour force, and 21 % of total exports, the Central Government implements agricultural resources development schemes under both central sector and centrally sponsored sector. These schemes generate voluminous information, both spatial and non-spatial, related to agricultural resources, using conventional, remote sensing and GPS technology.

As the Information Technology facilitates

(i) data support for better research, better results, and sustainable planning;

(ii) bridging the gap between knowledge and practice; and

(iii) judicious land use resource surveys, efficient management practices and sustainable use of natural resources.

It is required to develop "agricultural resources information system" using geomatics technology with public funding to reduce the process of marginalisation of small farmers and risks such as variation in output prices, etc., in India. This will facilitate to evolve "small farmer development strategy" making full use of frontier technologies (information technology and bio-technology) and ensuring the linkages between research, technology, and production on one hand, and effectiveness of the delivery system and extension network to carry the benefits of S&T to the farmers on the other [KVS2K].

In view of the increasing importance of Information Technology and Bio-Technology, the small farmers are compelled to :

(i) adapt their operations viz., group cooperation in the areas of land & water management, integrated pest management, integrated nutrient supply, and improved post-harvest technology;

(ii) organize themselves into "poly-centric" groups (to maintain and manage beneficiary programmes, planning, and evaluation), to achieve self sufficiency and an export surplus in farm production, and, in extreme cases,

(iii) resort to pooling of their lands into viable large "farming estates", avail of all latest technologies and compete in the global markets.

4

W.T.O. AND AGRICULTURAL TRADE

INTRODUCTION

More than ever before, developing countries are participating in and influencing international trade negotiations. GATT membership now includes 98 developing countries or economies in transition out of a total membership of 124. In 1948, the original 23 GATT contracting parties included only 11 developing countries. Furthermore, in the Uruguay Round (UR) of the GATT negotiations, which began in 1986, developing countries were able to work for the first time as a bloc.

Experts agree that developing countries have much at stake with the new UR agreement. Six studies have estimated the effects of the trade agreement worldwide, as well as on developed and developing countries. The estimates of worldwide income gain after implementation of the agreement range from $140 billion to $274.1 billion annually. Developing countries, in general, are expected to increase incomes from $36 billion to $89.1 billion annually, according to these studies.

While, in aggregate, the trade agreement is expected to benefit most countries, distribution will be uneven. Some regions, countries, or sectors within countries may lose, in the short run at least. Many trade analysts concur that those most likely to be negatively affected include: the poorest countries, net food importers, producers of coffee, cocoa, tea and rice, and countries that have previously benefited from preferential treatment. Sub-Saharan African countries are typically in these categories. At the same time, experts predict that the developing countries in Asia are going to be the greatest beneficiaries of trade reform among developing countries, with those in Latin America second.

Several issues in the agreement are of primary importance for developing countries. The World Trade Organization (WTO) will replace GATT as the mechanism to enforce the rules and settle disputes. Unlike in the past, countries must agree to virtually all the provisions of the agreement in order to be a member of the WTO. Agriculture, for the first time, will follow uniform trade rules under the WTO. The general reduction in agricultural subsidies is expected to improve market access and developing country competitiveness in a sector of importance to most developing countries. Another benefit to many developing countries from the UR agreement is the phase out of the Multi-Fiber Arrangement (MFA). Textile exporters, often developing countries, no longer will be subject to the discriminatory use of quotas that had been permitted by the MFA. Despite the many gains from the new trade rules, however, developing countries have some concerns: loss of market access due to erosion of preferential treatment, possible increase in food prices, loss of protection for domestic producers and consumers, and possible loss in value of certain exports.

The total impact from the UR agreement on developing countries will be evaluated throughout the next decade, and beyond, as it is implemented. Whether the UR agreement will affect US foreign aid, foreign investment or foreign policy may be a topic of hearings and debate in the years ahead.

GATT : THE URUGUAY ROUND AGREEMENT AND DEVELOPING COUNTRIES

On December 8, 1994, President Clinton signed the Uruguay

Round implementing legislation (P.L. 103-465), putting the provisions of the Uruguay Round trade agreement into law for the United States. Developed and developing countries alike will be greatly affected by implementation of the agreement. Some claim that it represents the most sweeping changes in world trade rules in 40 years.

How developing countries fare under the trade agreement is of some consequence to US foreign policy interests, how the United States implements its foreign aid program and longer term US trade and investment prospects. Developing countries are the fastest growing markets for US exports, and therefore, are gaining in economic importance. If developing countries expand their ability to trade and increase foreign exchange earnings as a result of the agreement, fewer demands for US foreign assistance could result. In addition, US trade and investment activities could be focused more in those developing country economies with upward trends. If, on the other hand, some developing countries are hurt by the new agreement, US foreign assistance and multilateral development aid may need to adjust their strategies to meet future needs. During the GATT negotiations, developing countries increased their political strength by operating as a uniform block of nations. This, coupled with their resources of certain strategic materials, for example, may be a factor in US foreign policy decision-making.

Included among the trade reforms are reductions in tariffs, an action that will likely result in greater market access for most exporting countries. Specifically, an expected increase in agriculture and textile market access could be of particular benefit to many developing countries. Rules for safeguards, antidumping, and subsidies are generally strengthened by the Uruguay Round (UR) agreement. Also, an end to voluntary export restraints (VERs), tightened rules for defining dumping, as well as a sunset provision for antidumping duties may reduce market uncertainty for developed and developing countries. A requirement to become a member of the World Trade Organization (WTO) is intended to prevent countries from being free riders, benefiting without making the commitment to abide by the rules. Each developed and developing WTO member country will have equal voting power in enforcing the agreement.

Recently, a number of economists, development experts and journalists have assessed the UR agreement affects on developing countries. There seems to be consensus among economists that most developing countries will be better off with the agreement than without it. Without the accord, these countries would fare far worse with the status quo, or under a more likely scenario of a trade war, experts believe. However, some journalists and humanitarian interest groups argue that the poorest countries (particularly in Africa) may be negatively affected, at least in the short run, by the new trade rules. This report presents the range of estimates, derived from the available studies, of world and developing country income gains or losses after implementation of the UR agreement. Included is a discussion on specific aspects of the UR agreement relevant to developing countries, as well as some of their concerns. Whether the UR agreement will affect US foreign aid, foreign investment or foreign policy may be a topic of hearings and debate in the years ahead.

BACKGROUND

The General Agreement on Tariffs and Trade (GATT) was first implemented in 1948 as a worldwide mechanism to promote free and fair trade among member countries. At that time, 23 countries, including the United States, signed on as contracting parties. Less than half (11) of the original contracting parties were developing countries, and thus, it was sometimes referred to as a "rich man's club." Today, there are 124 members, 98 of which are developing countries or economies in transition. Six additional developing countries and 14 transition economies are seeking to join the new World Trade Organization (WTO).

Several rounds of negotiations of trade rules have occurred throughout the history of GATT. The Uruguay Round, which began in 1986, is the eighth. On April 16,1994, officials from more than 100 countries gathered in Marrakech, Morocco to sign the UR agreement, also referred to as the Final Act. As of December 31, 1994, 80 countries have accepted the WTO agreement.

Estimates of Worldwide Income Changes Resulting from the Agreement

Experts agree that world income and the income of all three

country categories developed, developing and transitional economies are expected to increase after full implementation of the agreement. The estimated range of worldwide gain after implementation of the agreement extends from $140 billion to $274.1 billion. For developing countries income potentially will grow between $36 billion to $89.1 billion, according to these studies. Two estimates project developing country growth due to the UR agreement at 0.6 percent and 3 percent.

The studies also agree that world income gains or losses will determine how specific countries and exporters of specific products will do. Some analysts believe that as world income rises, so will demand for almost all products, and export earnings for virtually all countries will increase. Most economists believe that net food importers could suffer a loss (at least in the short run), although a few claim that even those countries could benefit. One study concludes that countries that export coffee, cocoa, tea and rice are least likely to increase exports, as prices for these products are expected to decline and demand is not likely to increase significantly.

Although useful for assessing how the GATT accord will affect the developing world, these studies have a number of limitations. For example, the quantitative studies lump developing countries together as one block. Development experts assert, however, that developing countries cannot be viewed as one single, homogeneous trading block. Each came to the negotiating table with its own problems and goals. These studies conclude that developing countries in aggregate will benefit from the agreement, but they acknowledge that the distribution of benefits will be uneven. Furthermore, some countries may gain, even though some of their population (usually the rural segment) may lose due to uneven distribution of income gains or imported food; in some cases, though, the urban population may lose if food prices rise dramatically.

An Agency for International Development (AID) report raises several concerns about quantitative assessments of developing country benefits from the UR agreement. First, some studies derive an estimate as the residual of the world minus industrial country gains, providing possibly misleading and overly optimistic estimates for developing countries. Second, while most studies conclude that developing countries generally will gain, they do not take into account the lack of resilience and leverage individual countries have in exerting influence on world supply, demand or price. Third, revenue gains for developing countries have not included the additional technical and financial resources that would be needed for a robust and rapid supply response to price or demand changes in the world. Many of the world's developing countries earn foreign exchange via exports of tree crops, the supply of which cannot be increased significantly in the short run. Furthermore, these studies have not considered the lack of infrastructure, agribusiness equipment, land restrictions or climate. And, finally, economic methods used to estimate income gains or losses from the UR agreement have been derived from known responses of first world economies, not third world ones.

An International Monetary Fund (IMF) report states that there are many problems with the quantitative assessments of the UR agreement which tend to understate the world and developing country gains from the agreement. First, precisely quantifying the benefits from changes in nontariff barriers (NTBs) is impossible; second, most of the studies have primarily an agricultural focus, thereby underestimating trade liberalization in the manufacturing and services sectors; third, the general equilibrium model assumed perfect competition in product markets, thereby excluding gains from trade due to economies of scale with imperfect competition; and fourth, the studies compared real income gains with the status quo, rather than with increased protectionism that many experts believe likely would occur without an agreement. Thus, the IMF conclusion is that the real income estimates of the UR studies are underestimated. A Brookings occasional paper acknowledges that estimates of world gain resulting from trade reform may be inaccurate, but suggests that even if the income gain resulting from the UR agreement is closer to its low-end estimate of $140 billion, that is still significant.

RESULTS OF THE AGREEMENT AND ISSUES FOR DEVELOPING COUNTRIES

With a greater level of participation by developing countries than ever before in GATT negotiations, these countries were able to attain a number of trade changes that will benefit them.

Increased market access, including agriculture, improved textile trade rules, membership and voting power in the new World Trade Organization, and strengthening of other trade rules are expected to help developing countries. Agriculture and textile reforms are potentially the most beneficial for developing countries. At the same time, some developing country governments have concerns about a possible gap in food availability and their food import needs, a weakening of preferential treatment, loss of domestic industry protection, and increased accountability for WTO membership.

TARIFF REDUCTION AND MARKET ACCESS

The agreement will reduce tariffs by an average of 38 percent on traded industrial products, with some tariffs being eliminated and some being reduced more than 50 percent. Tariff reductions will be in place after five years, with some product exceptions allowing for 10 year phase-in of tariff reductions. The UR agreement is the first to include significant agriculture and textile reform. Agricultural and textile tariffication (converting all NTBs to tariff equivalents) and tariff reduction may be among the most significant changes in world trade rules for developing countries.

Reducing tariffs could be a double-edged sword for developing countries. Lower worldwide tariffs generally will increase market access for their products, increase their potential to market their exports, and increase their foreign exchange earnings. Lowering their own barriers will likely mean an increase in consumer products at lower costs, as well. The negative side of reducing tariffs for many developing countries is that, particularly in Africa, developing countries have the highest tariff and nontariff barriers in the world. Reduction will mean less protection for domestic industries and could mean a significant loss of government tariff revenue in these countries.

A concern on the part of many countries (primarily ACP countries—African, Caribbean and Pacific) that have been recipients of preferential treatment (i.e., Generalized System of Preferences—GSP—by the United States and the Lome Convention by the European Union—EU) is that their preferential treatment in the form of lower or zero tariffs on certain products to certain markets will be diluted by significantly reducing tariffs to all under the UR agreement. Losses in ACP volume of exports to the EU market because of erosion of Lome benefits translates to increasing competition for African exporters with Latin American and Asian exporters of tropical products to the EU.

In contrast, the GATT Director-General disagrees with the concern that the erosion of preferential treatment will be devastating for ACP countries. The objective of GSP is not to divert trade from other exporters, but rather to provide the possibility for developing countries to compete on an equal footing with producers of industrial goods in developed importing countries. According to the GATT Director-General, this same objective can be effectively met through the elimination of tariffs negotiated in a multilateral trade agreement. He believes developing countries can compete with confidence about the future conditions of market access, rather than being affected by political philosophy or being conditional or temporary.

Some concern is currently being expressed by trade analysts as some countries appear to be binding their tariffs (including their tariffied NTBs) to levels that are suspected to be higher than they should be. Reducing them by the amount required by the agreement then places the protection higher (some reportedly are believed to be as much as 200 percent to 300 percent higher) than they were before the trade agreement. This practice is referred to as "dirty tariffication". Countries accused of doing this include the EU, the United States, Canada, Pakistan, Morocco, and Turkey, among others. Developing countries, especially, could lose much-needed benefits of market access (such as increased foreign exchange earnings from trade) if this practice becomes widespread.

AGRICULTURE

For the first time, the Uruguay Round Agreement will require agriculture to follow uniform world trade rules under GATT. Developed country agricultural tariffs (as well as converted NTBs) will be reduced by 36 percent, while those of developing countries will be reduced by 24 percent, using a 1986-1988 base. Each tariff item in developed countries must be cut by a minimum of 15 percent, and by at least 10 percent in developing countries. Internal agricultural support programs that distort trade will be cut by 20 percent, and export subsidy programs must be reduced by 36 percent in developed countries (over 6 years) and 21 percent in developing countries (over 10 years). Least developed countries are completely exempt from these agricultural reforms.

Developing countries have several concerns about the agriculture measures of the IJR agreement.

(1) Loss of Market Access due to Preferential Treatment

Preferential treatment has provided ACP countries with an advantage over other exporters by lowering or eliminating tariffs on some products into certain markets. ACP countries fear lowering tariffs worldwide to all exporters will dilute the advantage they previously maintained, thereby losing the marketing edge they have held in tropical products exports, particularly in Europe. It is possible that trade reform will help Latin American and Asian exporting countries to gain market access over the ACP countries in the near term. ACP losses in volume cannot be significant barriers to tropical goods that will decline because of trade reforms in the EU market.

(2) Possible Increase in Food Prices

Net food importing countries may find it increasingly difficult to pay for needed food imports as the UR agreement is expected to restrict world food supply. As food exporters are required to reduce internal supports and export subsidies, world food supplies are expected to decline initially. That, it is anticipated, will result in a rise in world food prices in the short run. Those countries that depend heavily on imported food or foreign food aid may have to use more of their scarce foreign exchange to purchase their current food import needs.

(3) Loss of Protection for Domestic Producers and Consumers

In many developing countries, consumers have received subsidies while producers have been taxed. Some countries have further imposed high tariffs or NTBs to restrict imports and protect domestic producers. The agreement will reverse these policies, allowing producers to react to market forces. In the long run most development experts believe that this will result in greater food security for developing countries. Nevertheless developing country governments are concerned about the transition period.

(4) Loss to Food Importing Countries

Countries that export coffee, cocoa, tea and rice could lose as the prices of their products are expected, by at least one study, to decline without a corresponding increase in demand. Many countries exporting these products are also net food importers. Being a least developed country and exempt from agricultural reforms may not help these countries. Goldin, Knudsen and Mensbrugghe believe that the major producers of these crops tend to tax the production of them. They argue that as prices for these commodities rise, domestic consumption declines. As producers increase exports, world price declines and receipts to these countries fall. Unfortunately, the exporters of these goods are often net food importers who will be faced with a price rise in staple foods such as grains and vegetable oils.

Many development experts and agricultural economists believe that if developing countries liberalize their own agricultural trade barriers, create an atmosphere that promotes foreign investment, and encourage labour to move out of unprofitable sectors into more profitable ones, developing countries would likely increase their own agricultural exports, increase foreign exchange earnings, and increase import purchases in the long run. Experts say that if world income rises, as it is expected to, demand will grow for virtually all exports, with developing countries being able to realize gains with increased demand from income effects. Moreover, some agricultural economists assert that world food price increases will encourage developing countries to boost their own food production, both to save on their food import expenditures and to increase foreign exchange earnings.

According to an AID study, "this will be good for domestic consumer, producer and country, if domestically produced food is nutritional and accessible to all at the same or lower costs as imports." There is a concern, however, that domestically-produced food will not be of adequate nutritional value and that the most rural populations in some developing countries will suffer from rising food prices and unequal distribution of food or income as a result of trade reform. The AID study goes on to criticize the theory that higher world food prices will benefit developing countries, saying that net food importers could be hurt, possibly in the long run, if they cannot develop increased food production capability. Those who believe that developing countries can gain from the UR by increasing their own export sales of agricultural commodities assume that these countries have additional technical and financial resources, as well as infrastructure, productive land, adequate climate for rapidly increasing food output. In many cases the exports of these countries are tree crops which would take a number of years to expand.

FOOD AID

The UR agreement specifically exempts food aid programs from required reforms. Bona fide food assistance, such as the U.S. P.L. 480's Titles I, II and III, as well as U.S. export credit guarantee programs are allowed to continue. The agreement also prohibits countries from circumventing required export subsidy reductions by moving commodities into food aid programs.

While bona fide food aid programs have not been directly challenged by the UR agreement, they will likely be affected if agricultural commodity supplies decrease and prices increase as a result of the agreement. Less purchasing power on the part of donor countries to buy and ship an adequate tonnage of food aid commodities could result in shortages to meet world food aid needs. Adding to the problem are increasing budget strains most food aid donor countries are experiencing, causing many developing countries to be worried about the future of food aid in the short run and possibly in the long run. Concerns prompted the food importing group of GATT participating in the Uruguay Round to add to the agreement ministerial declarations that specifically address these issues of net food importing countries, including a possible transition period for food prices and special cooperation with the international financial institutions.

Another view of food aid, however, is that developing countries have been hindered for years by the large-scale use of foreign food aid, allowing these governments to continue domestic policies such as consumer subsidies, and blocking needed market signals that would prompt developing country farmers to produce based on supply, demand and price. Some observers believe that limiting food aid only to emergencies (although the UR agreement does not do this) would improve food security in developing countries in the long run.

TEXTILES

The textile sector trade liberalization under the UR agreement is of keen interest to many developing countries. The Multi-Fiber Arrangement (MFA) has existed since 1974 and allows bilateral quotas to be established between textile exporting and importing countries, allowing for discrimination between them on a case-by-case basis. Under the agreement, the MFA will be phased out over ten years. MFA quotas have been applied mostly by developed countries against developing countries, although MFA has also allowed Pakistan and India to maintain high barriers. Thus, phasing it out and putting all textile exporters on equal terms favours developing countries, since they have inexpensive labour in this labour-intensive industry. Countries expected to gain the most from the textile provisions include China (not currently a GATT member), low-income countries in South Asia and the newly industrialized countries (NICs)—Hong Kong, Singapore, South Korea, and Taiwan. China is expected to gain in the medium term, although U.S. quotas for China's textile imports are not expected to be lifted until China becomes a member of the World Trade Organization (WTO). Sub-Saharan African countries are not expected to gain significantly because they already had been receiving preferential treatment for their exports to Europe, their primary market. (Although Mauritius and Nigeria have been shown to have potential gain with elimination of the MFA.)

A criticism of the UR agreement on textiles is that it is not until the seventh year that tariffs are reduced by 50 percent. The largest chunk of tariffs is not scheduled to be phased out until the tenth year. Furthermore, developed countries can offer the least domestically (politically or economically) sensitive quotas to lower tariffs in the early years, maintaining a continued higher level of protection for the last of the ten years.

WORLD TRADE ORGANISATION

With the implementation of the UR agreement, a new multilateral institution—the World Trade Organization (WTO)—was emerged to enforce the international trading rules, provide a unified mechanism for settling disputes, and provide the framework and flexibility for future world trade negotiations as new issues evolve. The WTO will also be responsible for examining member countries' trade policies and practices through the Trade Policy Review Mechanism. Membership in WTO will automatically mean accepting all the provisions of the UR agreement, for the most part. Therefore, countries will be prohibited from gaining benefits of the UR agreement without agreeing to comply with the rules, which a number of mostly developing countries had done under the old GATT. Some journalists and political action groups have expressed anxiety that WTO member countries, independent of their economic or political importance, will have equal voting power. Developing country governments believe that having equivalent voting strength with major economic powers such as the United States, Japan and Germany will expand their influence in world trade.

Prior to the U.S. congressional vote, consumer activist Ralph Nader suggested that a few small countries could ban together to block policies that large countries, such as the United States, would favour. Most experts, however, do not list this as a major concern regarding the WTO.

OTHER ISSUES

Other provisions in the new UR agreement will have some impact on developing countries, although less than the issues discussed above. Trade rules were strengthened by clarifying, defining, or setting deadlines. Of particular interest to developing countries are those changes concerning reform of safeguards, antidumping and countervailing duties, subsidies, state trading enterprises, customs valuation, sanitary and phytosanitary measures, preshipment inspection and rules of origin.

Safeguards allow a country to temporarily protect a domestic industry from serious injury due to sudden increased import competition. Voluntary export restraints (VERs), not sanctioned by the GATT, are bilaterally-negotiated agreements to restrict exports. Because of the growing use and abuse of this trade policy tool, the UR agreement abolished VERs. The agreement also changed the rules for use of antidumping measures and established a sunset provision to limit their duration. Tightening and clarifying trade rules and safeguards provides greater worldwide market stability and less uncertainty for developing country exporters.

The UR agreement, for the first time, defined subsidies as permitted subsidies, prohibited subsidies and an intermediate category—those that could lead to countervailing duties. Improved definition in these areas is expected to lead to a reduction in the use of countervailing duties. One area of concern among developing countries is that the new rules could result in an increase in government subsidies for research and development of high-tech products that they believe would continue a trade bias against poorer, developing countries.

Sanitary and phytosanitary (S&P) measures in the UR agreement require that scientific evidence be the basis for their use to restrict trade. The agreement allows use of the principle of equivalence—accepting standards that provide equal risk or protection, but may not be identical. Developing countries are concerned that S&P codes may require substantial investment in technology for some countries to comply to the new S&P codes. At the same time, some developing countries have applied S&P measures to protect their own domestic agricultural-related industries, encouraging inefficiencies. By removing unscientific S&P barriers, experts believe some developing countries will have access to an increased quality and quantity of food imports.

Trade-related investment measures (TRIMs), trade-related intellectual property rights (TRIPs) and services were new issues negotiated in the UR Agreement. The agreement establishes rules for trade-distorting investment measures and includes a list of prohibited activities such as local content and trade balancing requirements. Developed countries are to eliminate prohibited activities in two years, developing countries in five years and least developing countries in seven years. Some development experts claim that the new rules on investment will reduce the bargaining power of developing countries with large multinational corporations. However, the new rules also might make investment in developing countries more attractive if the rules make investment more secure.

TRIPs provides rules to fight international trade in counterfeit goods. It sets binding standards for protection of copyrights, trademarks, patents and trade secrets. In the long run, developing countries may realize greater access to foreign technology as intellectual property is increasingly protected. However, in the short run and in some cases, the cost to the developing country of imported protected products may increase at the same time that the country may be losing a "pirate industry". Among the most often accused countries for TRIPs violations are China, South Korea, Brazil, Singapore, Taiwan and Thailand.

REGIONAL ESTIMATES AND ISSUES

Most studies and experts agree that among the winners and losers, the poorest countries are likely to gain the least, if at all from the Uruguay Round agreement in the near future. According to ODI, the greatest beneficiaries of trade reform among developing countries are those in Asia, with Latin America second, and the poorest countries (primarily those in Africa) last. A number of experts contend that these countries could benefit in varying degrees, depending on how much they reform their own trade policies.

Asia

Many Asian countries will benefit from phasing out the MFA. ASEAN countries also will likely gain in increased market access for coffee, cocoa, cut flowers and spices. Indonesia and Malaysia are expected to notably increase their coffee exports. According to Goldin, Knudsen and Mensbrugghe, low income Asian countries' incomes are predicted to rise by 0.6 percent, although Indonesia's is expected to decline by 0.7 percent. The decline for Indonesia is due to the predicted declining prices of its major exports, such as rice, coffee and cocoa, while its large imports of wheat, meat and dairy are expected to increase in price.

China and India account for 40 percent of the world's population. Thus, how these two economies fare after trade reform is of interest to the rest of the world. Goldin, Knudsen, and Mensbrugghe estimate that China's income will increase by 2.5 percent after implementation of the UR agreement, while they estimate India's will increase by 0.5 percent.

In China, the gains will be observed more in the rural sector than the urban. With the exception of rice, most food prices are expected to rise as the cost of manufactured goods are expected to decline. Goldin, Knudsen and Mensbrugghe expect the world price of rice to decline due to the reduction of domestic taxation of production. At the same time, according to the study, consumer demand will decline because consumer subsidies will be reduced. Consumers likely will substitute other foods, such as animal and wheat products, which are not protected in China and are typically in greater demand as incomes rise.

India, which at first opposed large scale trade reforms (because of its high levels of protection), became a very active participant in the UR negotiations. (India realized that to avoid being shut out of international trade reform benefits altogether, it had to join the negotiations.) Nevertheless, many in India viewed the government's eventual cooperation as having sold out its own people, as it appeared to them that India was offering too many concessions. While India had already begun unilateral trade reform in 1991, cutting tariffs from the previous average of 125 percent to the current level of 70 percent, experts contend that it will have much further to go to meet GATT agreement requirements. Economists believe that India will gain significantly from the phasing out of the MFA and reform of agricultural trade rules.

Latin America

Generally, Latin America as a whole is expected to benefit most as a region from agricultural trade reform. Only one available study quantitatively assessed income gains in Latin America. Increases in exports of tropical products and meat are anticipated. Goldin, Knudsen and Mensbrugghe determined that Brazil would increase net income by 0.3 percent as of the year 2002; Mexico was shown to have no gain; and all other Latin American countries combined show a 0.6 percent income increase on an average. The subsequent rise in food prices is predicted to have a small net positive effect on much of Latin America, with the possible exception of Mexico and Brazil where the increase in food prices will not be offset by increases in food export earnings. Goldin, Knudsen and Mensbrugghe believe that Brazil will not fare as well as others with trade liberalization. Prices of Brazil's exports—coffee, cocoa, and orange juice—are expected to decline, thus resulting in a loss in farm income. At the same time, trade protection, which has been relatively high in Brazil (an average of 41 percent), will be reduced, causing a loss in rural and urban competitiveness that is not offset by lower domestic prices. Goldin, Knudsen and Mensbrugghe believe that Mexico will suffer a loss in export earnings which lowers national income. Both urban and rural sectors are likely to expand nominal income, but the country as a whole may sustain a loss as urban declining real income outweighs rural gains. The rural sector seems to benefit most with increases in dairy and livestock prices and lower prices for feedgrains.

Another concern on the part of ACP tropical products exporters is that with the reduction in tariffs, the preferential treatment they have received (largely exports of tropical products to Europe) will be diminished. Some believe that selected Latin American countries will increase the competition in European and U.S. markets for these products—a negative for ACP countries.

Africa

Many have highlighted African countries, generally, as the most likely to lose as a result of the UR agreement—from erosion of preferential treatment, expected increased food prices, and increased competition, as well as a loss of government revenue stemming from lower tariff collections. Some believe that Africa's economic future may be influenced more by structural adjustment and increased improvement of development programs than by the UR agreement. Nevertheless, liberalization of their own trade policies likely will determine whether these countries can realize any advantage as a result of the trade agreement, particularly since they maintain higher than average tariffs and heavily protect domestic production with trade barriers.

Northern African countries and South Africa have a greater potential to increase their income as a result of implementing and participating in the UR agreement and WTO. For the most part, sub-Saharan African countries were not actively involved in the GATT negotiations and are expected to be most negatively affected by them.

The agreement excludes least developed countries from many of its most stringent rules and allows for longer time periods to phase in most of the changes. Goldin, Knudsen and Mensbrugghe concluded that most African countries' incomes would decline after trade reform, but are optimistic that in the long run these countries could benefit if they made the required adjustments to liberalize their own trade rules. Goldin, Knudsen and Mensbrugghe estimate that trade reform similar to that of the UR agreement would result in an income loss for Nigeria of 0.4 percent and 0.2 percent for the rest of Africa, excluding South Africa, by the year 2002. This study estimates that South Africa will experience an increase in income of 0.6 percent by 2002.

According to Yeats, the erosion of Lome and GSP will result in an income loss estimated to be $4.2 billion for 29 sub-Saharan African countries. Cameroon, Cote d'Ivoire, Kenya, Senegal and Zimbabwe account for 85 percent of the $4.2 billion loss. Mauritius is expected to gain because of the textile agreement. Yeats predicts that sub-Saharan Africa will lose more exports through the erosion of preferential treatment than it will gain from trade liberalization.

ODI asserts that ACP losses in export volume to the EU market because of dilution of preferential treatment will not be offset by trade creation in other markets. The EU maintains most of the significant barriers that will be removed by the UR agreement. Thus, one of the primary markets to be opened by trade reform is the EU, already the primary market for ACP exports.

CONCLUSION

The general conclusion consistently arrived at by studies on developing countries and the Uruguay Round is that income will increase worldwide as trade expands. The Brookings study states that even the low-end estimates of worldwide income gain resulting from the agreement are significant.

The impact on developing countries in aggregate is expected to be positive, but uneven. Selected studies find that some—the least developed food importing countries—could be negatively affected in the short run. The Overseas Development Institute believes that even allowing for underestimation and restrictive assumptions, the total effect on developing countries is not likely to be massive. The studies agree that whether and how much a country benefits from the agreement depends directly on the extent of each country's own trade liberalization. Creating a favourable climate for foreign investment, allowing free movement of labour among industrial sectors, and lowering trade barriers which encourage inefficient production practices all will contribute to increased gains from the agreement.

The impact of the agreement on food aid, export credit programs and developing country food security can be fully understood only in this larger context. Whether trade reform has a significant impact of foreign food aid programs, U.S. foreign assistance, U.S. foreign policy, U.S. foreign trade and investment activities/programs may be a subject of hearings and debate as implementation of the agreement progresses.

Future issues likely to surface within the WTO framework include linkages between trade policy on the one hand, and environment and labour issues on the other. The newly established power of developing countries in WTO, however, could mean that these politically sensitive issues may be suppressed, despite the view of many international environment and labour experts that serious discussions need to take place.

EXTERNAL TRADE WITH INDIA

Agricultural exports from India were 44 percent of total exports in FY 1960; they decreased to 32 percent in FY 1970, to 31 percent in FY 1980, to 18.5 percent in FY 1988, and to 15.3 percent in FY 1993. This drop in agriculture's share was somewhat misleading because agricultural products, such as cotton and jute, that were exported in the raw form in the 1950s, have been exported as cotton yarn, fabrics, ready-made garments, coir yarn, and jute manufactures since the 1960s.

The composition of agricultural and allied products for export from India changed mainly because of the continuing growth of demand in the domestic market. This demand cut into the surplus available for export despite a continuing desire, on the part of government, to shore up the constant foreign-exchange shortage. In FY 1960, tea was the principal export by value. Oil cakes, tobacco, cashew kernels, spices, and raw cotton were about equal in value but were only one-eighth of the value of tea exports. By FY 1980, tea was still dominant, but coffee, rice, fish, and fish products came close, followed by oil cakes, cashew kernels, and cotton. In 1992-93 fish and fish products became the primary agricultural export, followed by oil meals, then cereals, and then tea. The share of fish products rose steadily from less than 2 percent of all agricultural exports in FY 1960, to 10 percent in FY 1980, to approximately 15 percent for the three-year period ending in FY 1990, and to 23 percent in FY 1992. The share of tea in agricultural exports fell from 40 percent in FY 1960 to roughly 17 percent in the FY 1988-FY 1990 period, and to only 13 percent by FY 1992.

EXTERNAL AID FOR INDIA

Foreign aid--financial and technical--since the 1950s has made a significant contribution to the agricultural progress in rural India. Aid has come from many sources: the United States government, the Ford Foundation, the Rockefeller Foundation, the World Bank, the Food and Agriculture Organization (FAO—see Glossary) of the United Nations (UN), the European Economic Community, the former Soviet Union, Britain, and Japan, among others .

Agricultural aid to India also has come in many forms. Between 1963 and 1972, for example, under a program of the United States Agency for International Development, some 400 American scientists and scholars served on the faculties of India's agricultural universities, while more than 500 faculty members from Indian institutions received advanced training in the United States and other countries. Several hundred agricultural research projects, financed with funds generated from sales of American farm commodities under the United States Public Law 480 program, fuelled technological breakthroughs in Indian agriculture.

Aid to the agricultural sector in India continued in the late 1980s and the early 1990s; the FAO, the European Union, the World Bank, and the United Nations Development Programme (UNDP) provided the bulk of the assistance. The FAO provided technical assistance in a number of emerging areas; it provided quality control for exports; videos for rural communication and training; and market studies for wool processing, mushroom production, and egg and poultry marketing. Operation Flood—a dairy development program—was jointly sponsored by the European Economic Community, the World Bank, and India's National Dairy Development Board. The UNDP provided technical assistance by sending foreign experts, consultants, and equipment to India. The World Bank and its affiliates supported agricultural extension, social (community-based) forestry, agricultural credit, dairy development, horticulture, seed development, rain-fed fish farms, storage, marketing, and irrigation.

India has not only been a receiver of aid. Increasingly since independence, India has been sharing its agricultural technology with other developing countries. Numerous foreign scientists have received special and advanced training in India; hundreds of foreign students have attended Indian state agricultural universities. Among other international agricultural endeavours, India has contributed scientists, services, and funds to the work of the International Rice Research Institute, headquartered in the Philippines. In the late 1980s and early 1990s, India provided short- and long-term training courses to hundreds of foreign specialists each year under a variety of programs, including the Technical Cooperation Scheme of the Colombo Plan for Cooperative Economic and Social Development in Asia and the Pacific and the Technical Cooperation Scheme of the Commonwealth of Nations Assistance Program.

IMPACT OF ECONOMIC REFORMS ON AGRICULTURE IN INDIA

The serious foreign-exchange crisis in 1990 led to a number of well-publicized economic reforms in the early 1990s dealing with trade, industrial licensing, and privatization. The reforms had an impact on the agricultural sector through the central government's effort to withdraw the fertilizer subsidy and place greater emphasis on agricultural exports. The cut in the fertilizer subsidy was a result of the government's commitment to reduce New Delhi's fiscal deficit by removing grants and subsidies from the budget. The government action led to a reduction in the use of chemical fertilizers and protests by farmers and opposition political parties. The government was forced to continue the subsidies but at a somewhat lower level.

New import and export policies aim at enhancing export capabilities of the agricultural sector by increasing productivity and promoting modernization and competitiveness. The measures to facilitate export growth include allowing the import of capital for the agricultural sector, reducing the list of agricultural products that cannot be exported, and removing the minimum export price from a number of products. Agricultural exports increased by 30 percent in FY 1991 and 14 percent in FY 1992 in terms of rupee value, but declined by 8 percent from FY 1990 to FY 1992 in United States dollar terms because of the devaluation of the rupee in 1991.

In the mid-1990s, it was expected that agriculture would continue to be the most important sector of the economy for the rest of the decade in terms of the proportion of GDP. However, even when it is not the sector providing the largest share of GDP, the importance of agriculture is not likely to diminish because of its critical role in providing food, wage goods, employment, and raw materials to industries. Despite their preoccupation with industrial development, India's planners and policy makers have had to acknowledge the critical role of agriculture in the early 1990s by changing basic policy. The gains in agricultural production should not lead to complacency, however. Continuing increases in productivity, developing allied activities in rural areas, and building infrastructure in rural areas are essential if India is to continue to be self-reliant in food and agricultural products and provide a modest surplus for exports.

AGRICULTURAL MARKETING IN INDIA

The agricultural marketing system in India operates primarily according to the forces of supply and demand in the private sector. Indian Government intervention is limited to protecting the interests of producers and consumers and promoting organized marketing of agricultural commodities. In 1991 there were 6,640 regulated markets to which the central government provided assistance in the establishment of infrastructure and in setting up rural warehouses. Various central government organizations are involved in agricultural marketing, including the Commission for Agricultural Costs and Prices, the Food Corporation of India, the Cotton Corporation of India, and the Jute Corporation of India. There also are specialized marketing boards for rubber, coffee, tea, tobacco, spices, coconut, oilseeds, vegetable oil, and horticulture.

A network of cooperatives at the local, state, and national levels assist in agricultural marketing in India. The major commodities handled are food grains, jute, cotton, sugar, milk, and areca nuts. Established in 1958 as the apex of the state marketing federations, the National Agricultural Cooperative Marketing Federation of India handles much of the domestic and most of the export marketing for its member organizations.

Large enterprises, such as cooperative Indian sugar factories, spinning mills, and solvent-extraction plants mostly handle their own marketing operations independently. Medium- and small-sized enterprises, such as rice mills, oil mills, cotton ginning and pressing units, and jute baling units, mostly are affiliated with cooperative marketing societies.

In the late 1980s, there were some 2,400 agro processing units in India in the cooperative sector. Of all the cooperative agroprocessing industries, cooperative sugar factories achieved the most notable success. The number of licensed or registered units remained at 232, of which 211 had been installed by March 1988. During the October 1987-September 1988 sugar season, 196 cooperative sugar factories were in production. They produced nearly 5.3 million tons of sugar, accounting for about 57.5 percent of the country's total production of 9.2 million tons. The National Federation of Cooperative Sugar Factories (India) rendered advice to member cooperatives on technical improvement, financial management, raw materials development, and inventory control.

In the early 1990s, the cooperative marketing structure comprised 6,777 primary marketing societies: 2,759 general-purpose societies at the mandi (wholesale markets in India) level and 4,018 special commodities societies for oilseeds and other such commodities. There were also 161 district or central societies covering nearly all important mandis in the country and twenty-nine general-purpose state cooperative marketing federations. The total value of agricultural produce marketed by cooperatives amounted to about Rs. 54.2 billion in FY 1988, compared with Rs. 18 billion in FY 1979. The total value of food grains handled by marketing cooperatives increased from Rs. 5 billion in FY 1979 to about Rs. 11.3 billion in FY 1986.

The Indian Ministry of Agriculture's Directorate of Marketing and Inspection is responsible for administering federal statutes concerned with the marketing of agricultural produce. Another function is market research. The directorate also works closely with states to provide agricultural marketing services that constitutionally come under state purview.

Under the Agricultural Produce (Grading and Marketing) Act of 1937, more than forty primary commodities are compulsorily graded for export and voluntarily graded for internal consumption. Although the regulation of commodity markets is a function of state government, the Directorate of Marketing and Inspection provides marketing and inspection services and financial aid down to the village level to help set up commodity grading centres in selected markets.

By the 1980s, warehouses for storing agricultural produce and farm supplies played an increasing role in government price support and price control programs and in distributing farm commodities and farm supplies. Because the public warehouses issue a receipt to the owners of stored goods on which loans can be raised, warehouses are also becoming important in agricultural finance. The Central Warehousing Corporation, an entity of the central government, operates warehouses at major points within its jurisdictions, and cooperatives operate warehouses in towns and villages. The growth of the warehousing system in India has resulted in a decline in weather damage to produce and in loss to rodents and other pests.

Most agricultural produce in India is sold by farmers in the private sector to moneylenders (to whom the farmer may be indebted) or to village traders. Produce is sold in various ways. It might be sold at a weekly village market in the farmer's own village or in a neighbouring village. If these outlets are not available, then produce might be sold at irregularly held markets in a nearby village or town, or in the mandi . Farmers also can sell to traders who come to the work site.

The Indian government has adopted various measures to improve agricultural marketing. These steps include establishing regulated markets, constructing warehouses, grading and standardizing produce, standardizing weights and measures, and providing information on agricultural prices over All India Radio (Akashvani), the national radio network.

The government's objective of providing reasonable prices for basic food commodities is achieved through the Public Distribution System, a network of 350,000 fair-price shops that are monitored by state governments. Channelling basic food commodities through the Public Distribution System serves as a conduit for reaching the truly needy and as a system for keeping general consumer prices in check. More than 80 percent of the supplies of grain to the Public Distribution System is provided by Punjab, Haryana, and western Uttar Pradesh.

The Food Corporation of India was established in 1965 as the public-sector marketing agency responsible for implementing government price policy through procurement and public distribution operations. It was intended to secure for the government a commanding position in the food-grain trade. By 1979 the corporation was operating in all states as the sole agent of the central government in food-grain procurement. The corporation uses the services of state government agencies and cooperatives in its operations.

The Food Corporation of India is the sole repository of food grains reserved for the Public Distribution System. Food grains, primarily wheat and rice, account for between 60 and 75 percent of the corporation's total annual purchases. Food-grain procurement was 8.9 million tons in FY 1971, 13.0 million tons in FY 1981, and 17.8 million tons in FY 1991. Food grains supplied through the Public Distribution System amounted to 7.8 million tons in FY 1971, 13.0 million tons in FY 1981, and 17.0 million tons in FY 1991, 21.2 million tons in 2008. The corporation has functioned effectively in providing price supports to farmers through its procurement scheme and in keeping a check on large price increases by providing food grains through the Public Distribution System.

5

Food Shortage : Meaning, Causes and Factors

INTRODUCTION

Food shortage occurs when food supplies within a bounded region do not provide the energy and nutrients needed by that region's population. Food shortage is most easily conceptualized as a production problem - not enough food is grown to meet regional needs - but constraints on importation as well as storage can also cause or contribute to food shortage. Food shortage is also created where food is exported from areas where production is adequate or even abundant. Historically, the great hunger of Ireland (1845-1847) and the famine of Bengal (1944) have been attributed more to British political decisions to export locally produced grain supplies without compensating imports than to production shortfalls per se (Woodham-Smith 1962; Sen 1981).

Even when production shortfall is the primary cause of insufficient supply, the ecological and political reasons for production problems vary widely. They range from natural disasters such as drought, flood, or fungus, to political disasters such as civil conflict, to misguided economic policies such as price controls- all of which discourage production of essential foods.

In all situations of food shortage, many within the region's population are hungry; but in every food-short region, others still enjoy adequate access to food. Equally, although many are food secure in areas of adequate food production, some still go hungry. These variable patterns of hunger result not only from skewed food distribution within regions based on differential political and economic resources, but also from selective marketing, and from non-market political policies of food extraction or assistance.

This chapter begins with an overview of the evidence for global and regional food shortage and where it is most likely to occur. It then analyses the causes of country or within-country food shortages. Finally, it considers the relationship between drought and famine, using recent evidence to argue that food shortage is not inevitable even in areas of widespread production failures: political, not environmental factors are the primary causes of food shortages.

IS THERE A WORLD FOOD SHORTAGE?

World agriculture produces enough food calories to meet the energy needs of all the nearly 6 billion (6 x 109) people who are alive today. Increased production based on advances in seed, water, and environmental technologies, and their wider dissemination especially in developing countries, have removed insufficient production as a cause of food shortage for the world as a whole. Global agriculture has managed to keep pace with population growth, and world food security is also safeguarded by cereal carry-over stocks; 19-20 per cent of annual cereal consumption is carried over into the next year to provide food in case of disastrous production failure (FAO 1993). Nevertheless, during any year in which enough calories are produced on a global level to meet the energy requirements of the entire population, food shortages can still occur under two situations. If the patterning of production directs too many calories into animals instead of humans, some enjoy meat while others lack calories. Alternatively, overemphasis on production of calories may jeopardize the production of other protein- or micronutrient-rich foods that also enter into the calculus of global food security or shortage. Both are production as well as distribution issues.

Dietary Factors in World Food Shortage

The numbers of people potentially supported by the global food supply depend heavily on the kind of diet people consume. The World Hunger Program calculates that global food supplies have been more than adequate, since the mid-1970s, to support the world's population on a vegetarian diet. But they would support only 74 per cent of the 1993 population on a diet where 15 per cent of calories come from animal foods (Uvin 1996). Only 56 per cent of the 1993 world population could have been provided with diets where 25 per cent of calories came from animal foods (Uvin 1996).

Vegetarian diets are typical in a wide range of developing countries, but worldwide the demand for meat is growing. Diets in industrialized countries differ widely with respect to their composition: those living in the United Kingdom eat less meat per capita than residents of the United States; meat consumption in Sweden is only 60 per cent of the US average (Bender 1994). Beef production continues to increase and poultry production is increasing faster than population growth rates. Although the *rate* of growth in the production of animal foods has been slower since 1980 than in the previous few decades, production of these foods continues to increase (Wiener and Wang 1990). Food shortage of the future, calculated on the basis of total future demand for grain consumed directly or in the form of animal foods, will be conditioned by whether peoples adopting richer diets follow the European or US pathway.

Food security must also take into account the growing demand for higher-cost cereals such as rice and wheat over sorghum and root-crops, as well as animal products. All are more resource expensive to produce. Increased demand for meat is a particular concern, since livestock conversions, usually calculated in terms of food energy grain-to-livestock ratios, are high. In a feedlot, it takes two kilos (kilograms) of grain to produce one kilo of chicken or fish, four kilos to produce one kilo of pork, and seven to produce one kilo of beef. Some suggest the ratios may be even higher: 3:1, 6:1, and 16:1. In the 1990s, it was calculated that some 4.3 billion large domesticated animals and 17 billion poultry eat 40 per cent of the world's grain supply (Foster 1992). Animal production also takes land and water resources. The argument can be made that these might be allocated otherwise to less resource-expensive food crops, but the livestock economy is complex, and reductions in animal production will not produce more food suddenly at lower cost.

Food shortage also coincides with adequate calorie production where the foods consumed are deficient in protein or micronutrients. Diets may be adequate in quantity but not quality. The three most common micronutrient deficiencies, those of iron, iodine, and vitamin A. Such deficiencies are most common in vegetarian or near-vegetarian diets that lack variety, both because such diets tend to be consumed by people who lack resources to acquire greater variety and because some nutrients are more abundant in animal foods.

Strategies to Change Consumption Patterns

These failures of the global food system to prevent food shortage despite adequate food energy production are being addressed in various ways in both developed and developing countries. In developed countries, higher consumption of animal foods is being discouraged. Meat consumption has begun to decline in response to health concerns that diets high in animal fat contribute to cardiovascular disease and certain cancers. Nutritionists are also trying to improve the dietary habits of poorer people, who tend to consume diets higher in fat. For reasons of equity and conservation, "food first," ecology, and natural resource advocates in developed countries also stress the importance of "eating low on the food chain"—more grain and vegetables, less meat—to make more food available to meet global food needs, and to encourage •sustainable agricultural practices (Lappe 1991; Brown and Kane 1994).

To meet world food needs, fat consumption also needs to be discouraged in some developing countries where average meat consumption is still at relatively healthy levels, but demands for animal protein and fats are increasing. Increasing the consumption of grains, legumes, root crops, and vegetables, relative to animal products, involves changing dietary patterns in developed countries and reversing current trends toward higher consumption of meat and fats in developing countries. Since higher status conventionally is associated with greater meat and fat consumption, such dietary change challenges traditional social, as well as nutritional, beliefs and practices and may prove difficult.

Micronutrient deficiencies, by contrast, can be targeted by a number of specific interventions: (1) increasing intake of foods containing them; (2) fortifying or enriching other foods so that they contain more of the needed nutrients, or (3) providing oral or injectable vitamins and minerals. To be successful, however, all rely on some degree of political commitment, as well as behavioural change. Although costs for some programmes are relatively low (e.g. salt iodization costs only about five cents per person per year, and vitamin A capsules only about six cents [Grant 1995]), they demand technical and social organization, and political and community support, for effective implementation and monitoring.

Despite these concerns about dietary quality, the most important conclusion to be drawn from an analysis of global food shortage is that there is no such shortage. Hunger is not primarily caused by food shortage.

WHERE ARE THERE REGIONAL FOOD SHORTAGES?

Although global food production has kept pace with world population growth, the rate of population growth has outstripped the rate of growth in food production significantly in some developing regions, and caused per capita food availability in these regions to decline. Per capita food production in developing countries overall has increased less than 10 per cent since 1960, despite impressive continued growth in *total* food production. Trends in food production are the worst in sub-Saharan Africa (SSA), where per capita food production has decreased slowly but relentlessly in recent years. Developed countries over this same period were able to increase both food production and production per capita, however, benefiting from advancing technology but also much lower population growth rates than in developing countries.

Per capita food availability has remained relatively constant in food-deficit regions and countries, primarily owing to food imports acquired through trade or aid (ACC/SCN 1993). FAO estimates of dietary energy supply (DES), which not only measure net food production (minus exports) and imports but also account for food lost, stored, or used for animal feed or industry, show that in only two regions—SSA and South Asia—does DES per capita fall below basic average requirements (set at 2250 calories). High fertility combined with underdeveloped agricultural technology and infrastructure, plus high incidence of natural disasters and civil disorder, make SSA the region most evidently food short. In South Asia, even though birth rates and population growth have slowed and agricultural output has increased dramatically in South Asia over the past two decades, there is still not enough food to ensure adequate nourishment for everyone.

Table 3.2
Per Capita Dietary Energy Supply by Region

Region	*Calories*	*Region*	*Calories*
Sub-Saharan Africa	2,099	Near East and North Africa	3,094
South Asia	2,245	Oceania	3,330
South-East Asia	2,446	Former USSR	3,380
South America	2,625	Europe	3,450
China	2,657	North America	3,600
Central America	2,822		

Two lessons that can be drawn from these low-DES regions are that severe production shortfalls are not so easily made up by food imports, and that population growth is not the only cause of food shortage.

In SSA, low DES is often attributed to low import capacity due to low export earnings and large burden of debt service. But even were these constraints removed, there would remain those of low economic (including agricultural) productivity, civil strife, and lack of infrastructure. Greater dependence on imports within the region would also magnify associated logistical problems—higher costs per unit food as demand increases, transportation costs to reach remoter areas, expanded costs of shipping, and storage losses. Multiplying numbers of conflicts and their aftermaths already have created unmet demand for food aid, especially as the costs of the actual food increase in an environment of shrinking foreign assistance.

In South Asia, the data indicate that food shortage remains a key issue, even where per capita food production has continued to rise along with population increase. Per capita DES has increased steadily in South Asia since 1970, despite very modest gains in per capita income (Uvin 1994). Even if lower caloric requirements are accepted for South Asians, based on their smaller body size, there are further signs of food shortage in this region.

Additional indicators of food shortage in both of these regions are provided by the World Food Programme, which has carried out far more emergency operations in SSA than elsewhere; their next most heavily aided region is Asia (Uvin 1994). Emergencies refer to both man-made and natural disasters. Furthermore, in these two regions, each of the three major micronutrient deficiency diseases is a serious public health problem, as contrasted with other regions that suffer from one or two but not all. Their combined food and nutritional deficiencies contribute to a continuing cycle of low productivity and hunger in the countries of these regions.

How Common are Country-level Food Shortages?

Global and even regional food availability estimates hide important variations in DES and food self-sufficiency of individual countries. Although Latin America as a whole has more than enough food to feed the region's population, seven countries in the region (with populations totalling 67.2 million people) had DES below requirement over the period from 1988 to 1990 (UNDP 1994). There are an additional 13.6 million people living in countries with inadequate food availability, despite being in regions where food supply is adequate. The remaining 721.5 million people living in hungry countries also live in hungry regions (Uvin 1996).

Food self-sufficiency is an additional issue and a potential determinant of food shortage. "Food First" advocates have argued that the only way for a country to prevent hunger is to promote food self-sufficiency or self-reliance (countries are able to trade for quantities sufficient to meet home country needs [Lappe and Collins 1978]). In their analysis, the country that lacks food self-sufficiency is a hungry country or, in our current terms, food short. Fortunately, the data do not support this oversimplification. Of the 99 countries that did not produce enough food to meet the needs of their national populations in the 1980s, only 48 (32 in SSA) were food short as measured by per capita DES (Uvin 1994). Thus, a country's dependence on imported food does not necessarily mean that more people in the country are hungry or that the country has exhausted its agricultural potential. Small industrial food-importing countries easily produce enough other goods to cover the costs of their food needs, purchased on the world market.

However, Lappe and Collins's assessment may be correct for countries with predominantly poor rural agricultural economies. The failure of a city-state such as Hong Kong to grow enough to feed its urban population has hunger implications that differ greatly from the failures of predominantly rural African countries whose export earnings are low. For poor agricultural populations, whose entitlements to food may come in large part from home production, their country's deteriorating position of food self-sufficiency may be an indicator of their own reduced access to food and resources to produce food.

CAUSES OF SHORTAGE

The discussion above on prevalence and indicators of food shortage has illustrated that its causes are complex. Some hunger indicators, such as production shortfalls, highlight problems that may lead to food shortage. Others, such as DES, directly measure food availability within a country or region. These food-shortage indicators report outcomes of physical and biological factors, sociocultural influences, political-economic forces, and interactions among these elements.

PHYSICAL AND BIOLOGICAL FACTORS

Production is only one determinant of food shortage, but a crucial one. It is obviously critical on the global level but can also be decisive for some countries or communities. Production potential varies across countries, dependent on natural factors (including climate, soils, water, food species, and pests) and cultural factors (including technology and investment strategies).

Climate

Temperature and rainfall are critical elements determining when and how often crops can be sown. While some Asian countries are able to harvest three times in a single year, food production nearly halts during dry seasons in many tropical zones and during winter cold in temperate areas. Extremes or thresholds of heat, increasingly accompanied by high ultraviolet radiation, and of cold, especially early frosts or late thaws, can ruin harvests. They test the limits of growing seasons and moisture-temperature tolerances of particular crop varieties. These extremes will be modified by global climate change, which promises to transform regional cropping patterns. For the present, drought is the most widespread climatic threat to production, and is treated more extensively below.

Seasonality means that there may be food shortage during part of the year in places where total annual production appears to be more than sufficient to meet nutritional needs. Agricultural societies and households adapt to potential seasonal scarcities by planting a variety of early-to later-yielding crops, storing or selling harvests to minimize losses, investing in social feasting when food is plentiful, and drawing on social obligations of reciprocity when food is scarce. They ration, process food staples more coarsely, supplement diets through foraging, and consume less-preferred foods. They also schedule crafts, migrant labour, and other economic activities to diversify and ensure income in "off" agricultural seasons. Despite such adaptive mechanisms, prolonged or multiple years of shortage, as experienced especially in SSA and South Asia, give rise to potential famine conditions (defined as widespread and extreme food shortage leading to elevated mortality and mass movements of population in search of food) that nowadays are addressed relatively successfully by state and international early warning systems (EWS) and response. Even where EWS are well established (as in India), however, seasonal hunger remains a problem addressed neither by state nor by traditional sociocultural mechanisms of food sharing, which tend to be undermined by modernization (Chen 1991).

Although drought is often thought of as the precipitating cause of famine, because so many farmers in a single area experience crop failure simultaneously, drought does not lead to food shortage or its extreme manifestation—famine—if there are adequate carry-over stocks available, or if food is available through market or relief channels (Ravallion 1987).

Creeping disasters, such as drought, are likely to devastate crops but leave infrastructure intact. Both timing and duration of rainfall may be implicated, since, for seed crops, moisture is critical immediately after planting and also at the stages when fertilizer is applied. Drought can also prevent planting if the rains are late, so that soils are too dry to till before the planting season has passed. Failure of rains at any of these stages greatly reduces harvests.

Since the early origins of agriculture, human societies have always tried to extend productive seasons and, especially, available moisture by controlling groundwater evapo-transpiration, by various water-storage techniques, and by distribution via irrigation. Water control is closely linked to social power and control, since the ability of irrigation systems to enhance food production is limited by their general state of repair. Hydraulic systems, which initially require large amounts of capital and labour to construct, later require maintenance. Social resources must be mobilized through a sense of common purpose (or coercion) to preserve irrigation ways, which otherwise silt up, leak, and lose effectiveness. The history of political fortunes and social breakdown in the Near East and Middle East has been linked to cycles of "salt and silt," environmental and subsistence crises triggered by failures to control the life-giving waters of irrigation (Jacobsen and Adams 1955). The expansion and sustainability of Green Revolution (GR) agriculture in the modern era is similarly dependent on effective water management, which is necessary to prevent waterlogging and salinization of soils and crops, and silting and disruption of water channels.

At the opposite end of the rainfall spectrum, too much rain can also impair agricultural production, especially where flooding is severe enough to kill crops by uprooting or submergence, but also where it simply slows growths or makes cultivation and harvesting extremely difficult. Even after harvest, flooding can be devastating when it occurs before crops have been safely stored (Good 1986) and leaves crops vulnerable to rot.

Irrigation systems can be overloaded by flooding but also can be one of the mechanisms used to cope with irregular rainfall. Irrigation systems often do little to prevent damage caused by the force of the rains themselves but they do prevent further damage from waterlogging. They also enhance productivity in dry seasons by holding over water from heavy rains; this is accomplished through the use of simple earthen dams in Sri Lanka (Grigg 1985), as well as through more elaborate systems of canals and pumps. Along with terracing, irrigation systems may also limit soil erosion and help sustain soil fertility along flood plains.

Less regularly, food systems can be entirely disrupted by natural disasters, which affect both growing and marketing conditions and sometimes the social fabric. Hurricanes and, to a lesser extent, earthquakes can destroy crops. They can also devastate transport, markets, and other infrastructure, and cause food shortage even where the crops themselves survive. The economic destruction accompanying natural disasters, which can cause scarcities of many materials, often pushes prices upward, increasing the rate of inflation, reducing employment, and more generally contributing to balance-of-payments problems.

Inability of government to deal effectively with natural disaster, such as earthquake or drought, can threaten political stability and overturn fragile regimes. In each of the recent cases of Ethiopia, the Sudan, and Rwanda, their ruling governments' ineffectiveness in dealing with drought and ensuing famine conditions provided an opening for the political opposition to challenge successfully that government's authority and legitimacy. Political leaders were portrayed as prospering while the masses went hungry, and this became the successful rallying point for civil uprisings. Similarly, the inability of Samoza's regime in Nicaragua to alleviate widespread suffering from a hurricane became the trigger cause of the Sandanista's successful ascent to political power. Each case, however, was followed by periods of readjustment, civil disorder, and lowered food productivity.

Although the magnitude of disasters' effects on overall economic development appears to vary (Albala-Bertrand 1993), the short-term economic shocks caused by sudden natural disasters usually decrease marketed food availability in the affected regions, so that food aid may appear to be the only practical means of getting enough food to the affected populations. Inappropriate world response, such as Guatemala's inundation with donated food following the 1976 earthquake, sometimes disrupts markets and income for local farmers, who face plummeting food prices and agricultural income even where their crops have not been destroyed.

Politicians and national politics are also important social actors in other ways. They provide the economic, social, and cultural framework (what policy makers increasingly term "the enabling environment") to prevent natural elements from precipitating wider disasters. Despite the seemingly arbitrary nature of sudden natural disasters, they do not affect all areas in the same way, even when their severity is of a similar scale. Healthy economies rapidly bounce back from shocks because they have more internal resources dedicated to mitigating the immediate and longer-term impacts on food production or distribution. Similarly, precautions that minimize the impact of disasters, such as earthquake-resistant housing and roads, are not distributed evenly either across or within countries. Some so-called "natural" disasters might be better thought of as man made. For example, flooding often results as much from deforestation as from excessive rainfall or unusually intense hurricanes. Abuse of land, particularly overgrazing, increases vulnerability to wind and flood erosion (Albala-Bertrand 1993). Sudden natural disasters cannot be prevented, but the effects that they have on food production and food importation are conditioned more by political and economic processes than by the intensity of the calamity.

Soils

Food production also varies according to soil structure and fertility, factors that are less easily measured—but perhaps more easily modified—than temperature or rainfall. The differences in productivity between the dark soils of the US midwest and the sands of the Sahara appear obvious, but it is less clear how much less-dramatic variations in soil conditions matter. Many tropical soils contain less nitrogen and phosphorus, have lower capacity to absorb fertilizers, and therefore have lower conventional productive capacity, but some tropical soils (most notably in the Amazon) have been very intensively farmed and further intensification is possible in other areas (NAS 1986).

Since agricultural methods and inputs vary in areas with different soils, the causes of disparities in productivity are multiple. Grain yields per hectare in SSA are about one-third of those achieved in East Asia, but SSA also struggles with more challenging climate, uses fertilizer at a rate less than 13 per cent of the world average (World Bank 1992), and has made little use of irrigation. Investments in agricultural intensification, including higher-yield-potential seeds, fertilizers, water management, and chemicals for pest control, are costly and make it unlikely that they will be easily or widely available for use by poorer farmers and countries. Especially where imported food is cheaper than domestically produced food, as is the case today in many developing countries, expanding local production may not appear to be economically feasible.

Production may also be limited by low soil fertility and restricted access to fertilizer supplies. Fertilizer is required to realize the full benefits from hybrid GR seeds; quantities of its use often serve as a measure of agricultural improvement or modernization. In contrast to the time when most farmers used local sources of animal or green manure to enhance soil fertility, most developing countries today rely heavily on inorganic fertilizers, which they must import. At both household and country levels, fertilizer constitutes a significant expense, and lack of means to purchase adequate quantities potentially reduces crop yields. Lack of cash, poor credit, or isolation from sources of supply due to underdeveloped infrastructure hamper farmers' access; lack of foreign exchange and balance-of-payments difficulties limit the total supplies imported and available within a country. The cost per unit of fertilizer is also a factor adding to the gap between fertilizer needs and the amount that many developing countries can afford to import (Monteón 1982; UNCTAD and Mukherjee 1985). Upward fluctuations in the price of petroleum, a raw material for inorganic fertilizers, imperilled developing country food production in the mid-1970s, and could happen again. Timely, as well as total, availability constrains output, since optimal fertilizer impact is achieved only by applying it at the most sensitive points in the crop cycle. Only greater domestic production can protect farmers and consumers in developing countries from severe price fluctuations and transport or market bottlenecks.

BIOLOGICAL STRESSORS

Disease, insects, animals, and weeds also damage crops and reduce yields, and are controlled by mechanical, land management, chemical, or biological means, including breeding plants and animals to resist key stressors.

Viruses are controlled by eliminating insect or other vectors and by breeding resistant seeds through conventional or new genetic engineering techniques. Biotechnology also offers new diagnostic techniques for identifying and limiting infections, as well as for producing and multiplying clean seeding stocks for vegetatively propagated crops, such as potatoes and manioc. Diagnostics and planting-material multiplication potentially can be carried out as local-level cottage industries. Bacterial and fungal diseases are also addressed through breeding programmes and sometimes chemical applications. In addition to the breeding and selection of resistant stock, animal diseases are controlled by preventive inoculations and curative remedies. Biological and chemical control of vectors are also common, and sometimes innovative, as where scientists of the Kenya-based International Centre of Insect Physiology and Ecology developed a simple cow urine-baited trap for tsetse flies, to eliminate trypanosomes.

Insect damage similarly is avoided by breeding resistant varieties, mixed-planting strategies that buffer particular host species from their characteristic pests, and insecticidal chemicals that kill (or otherwise interfere with the feeding, maturation, or sexual reproduction of) the target populations. Biological controls involve introducing predators or pathogens of the pest species. Integrated pest management that combines chemical, biological, and some hands-on mechanical strategies - such as removal of insect eggs before they hatch - are increasing worldwide, in response in part to their greater safety and lower costs and in part to the increasing environmental burden and decreasing effectiveness of chemicals. Chemical use soared during the early decades of the GR, as new seeds were accompanied by chemical packages that sometimes indiscriminately wiped out "good" as well as "bad" insects, removing the predators as well as pests, presenting an unprecedented opportunity for pesticide-resistant insects that co-evolved with the plants and chemicals in these GR ecosystems to cause severe crop damage. To restore insect ecology, reduce pesticide poisoning, and re-create a safer balance for plants and humans, Indonesia, as a case in point, banned most pesticides. On Java, in some seasons and places where rice-hopper damage can be anticipated to be enormous, communities have organized massive brigades of farmers and schoolchildren to collect the insects' eggs and so prevent damage (Indonesian National IPM Program 1991).

Other kinds of animal damage, such as that of rodents, birds, or livestock, are conventionally limited by human labour, although the "excess" child labour that traditionally scared birds and livestock from ripening fields is disappearing as schooling competes with agricultural tasks. Reduced child labour availability is one of the factors leading to the selection of maize over sorghum in many parts of Africa, since maize ears are covered and less subject to damage than sorghum seeds, which are exposed. Rodent pests may be chased, physically killed, or poisoned, although the poisons are still not well controlled. Human poisoning is a hazard in cases of insecticides and rodent poisons, since the toxic chemicals tend to be sold in small quantities and not handled with the care specified by their manufacturers. The usual way to prevent larger animals stealing crops is scrupulous tending in areas where such risks are high.

Weed damage also is increasing, as plagues such as Striga spread across the maize and sorghum fields of Africa. Weeds choke out useful plants and also compete with cultivated plantings for soil nutrients and water, reducing their yields and disease/insect resistance. Careful plantings, to give the intended food crop a healthy start, or herbicides are alternative strategies, as less labour is available to hand-weed at very low wages.

Seeds of major crops that have been genetically engineered to resist a particular brand of herbicide also are being aggressively developed and marketed by large chemical giants, such as Monsanto, that have purchased seed companies and the molecular-biologist/plant-breeding scientists to advance the match and ensure a continuing market for their chemicals. These developments, although they may increase crop production, also increase the chemical load and the potential for herbicide-resistant genes (traits) to pass from the desirable species to closely related weed species, which would then be equipped to inflict even more damage.

Two modern agricultural factors contributing to greater risk of pest damage are genetic uniformity and uniform or continuous cropping cycles. As most farmers in a region purchase the same single varieties of improved seeds designed to maximize production on the same agricultural cycles, they establish conditions ideal for the explosion of many pests. An overall strategy to reduce plant damage therefore is diversified cropping. Carefully patterned plantings of multiple species and varieties within species is a good hedge and an aspect of traditional agricultural practices that modern cultivators need to review.

Seeds and Related Technologies

Basic to all production strategies and yields are the seeds, which may be more or less productive; drought and disease resistant; or responsive to moisture, fertilizer, and other chemical inputs. In many regions, "traditional" seed varieties are less and less plentiful because they are unable to keep up with demand for more food and more intensive methods. Modern seeds, bred for shorter growing periods and neutral photoperiod, allow multiple sowings and harvests; they also may be tailored to resist insects and plagues that attack traditional varieties and lower their yields.

Exponential growth in world food production over the past three decades has been made possible largely by new seeds and related technological advances that have greatly increased output per unit of land and labour. The impact of the GR seed-water-chemical technologies on grain yields has greatly reduced food shortage, especially in rice- and wheat-growing regions.

But technological advance can prevent food shortage only if: (1) it continues to increase yields faster than the population grows, (2) food availability within agriculturally modernizing regions actually increases, and (3) increased caloric availability is not transformed into richer diets.

Controversy rages over whether the dramatic food production increases achieved over 1960-1990 are sustainable and will be able to keep up with projected population growth. Brown and Kane (1994), who think that world population has reached its agricultural limits, argue that prior to the GR there existed a backlog of agricultural technologies waiting to be applied. Today, by contrast, no such "breakthrough" technologies are waiting "on the shelf"; although advancing biotechnologies promise to enhance production, they will not allow the quantum jumps in production necessary to keep pace with growing population. Technological optimists, such as World Bank economists Mitchell and Ingco (1993), differ: they argue that biotechnologies hold very great potential to increase food supplies and that careful pricing of both factors of production and resulting food products also can stimulate more food production. Intermediate positions, such as that of the International Food Policy Research Institute (IFPRI) "2020 Vision for Food, Agriculture, and Environment," argue that there also exists greater potential than Brown and Kane allow, to expand the applications of known technologies to prevent regional shortages, but that the world community must make such agricultural investments their priority. Because of required investments and economies of scale, many poorer farmers in developing countries have not been able to take full advantage of existing technologies. Increasing food production in some areas may depend more on altering social and institutional structures to allow fuller utilization of existing technology; it depends less on radical new breakthroughs.

Other intermediate positions suggest that there exists great potential for significant reductions in expenditures, and increases in available food supplies, by controlling waste all through the food system (Bender 1994). Post-harvest losses in cereals are coming under control (Good 1986; Donahaye and Messer 1992) but those of roots, tubers, fruits, and vegetables remain very high: losses of potatoes are estimated at 5-40 per cent and of bananas at 2080 per cent (Tome et al. 1991). Biotechnology can also contribute to reductions in loss over the entire cultivation cycle and after harvest (Donahaye and Messer 1992).

Such savings and incremental production will not automatically eliminate food shortage for hungry populations, however. Technological advances still need to be directed toward crops and cropping characteristics, and distributed in manners that benefit those vulnerable to hunger. A case in point is Mexico, which has continued to suffer food shortage despite more than doubling its food production over 1960-1985, a period over which the volume of imported grain increased more than 20-fold (Barkin et al. 1990). This massive increase in imports, concurrent with rapid expansion of domestic production, can be accounted for in part by the shift of land out of maize production and into sorghum. Although sorghum is a basic food crop consumed in many areas of the world, in Mexico it is used primarily for animal food and therefore actually contributed to domestic food shortage because of the associated drop in maize production and export of livestock.

In sum, food production, supply, and the productive environment are influenced by a range of physical and biological factors that are not easily separable from sociocultural and political-economic conditions. These illustrate the linkages between production and consumption at local, country, and global levels.

SOCIAL AND CULTURAL FACTORS

The causes of food shortage are in no way limited to physical and biological factors affecting production. Sociocultural factors include the organization of land and labour use as well as dietary preferences. Political-economic factors involve world markets and government policies to modernize agriculture and increase foreign exchange, which at the local level are translated into incentive structures, especially "getting the prices right."

Land and Water Control

Land tenure greatly influences what and how much is grown. Motivations to invest in land improvements and agricultural technology are tied to anticipated returns from land ownership or rents. Land reform is often cited as the major policy tool to improve food and other agricultural output, but may be most difficult to implement, since it threatens vested interests. Barraclaugh (1989), in his probing analysis of policies to end hunger, found that even allegedly reformist governments, such as the Sandinistas in Nicaragua, are stymied by vested landholding power interests if they seek more equitable land allocations. Communal or private ownership, permanent or usufruct use rights, and tax or labour obligations associated with tenure are three kinds of conditions that define the social and political order and the social relations of production and consumption. Inequality in landholding is an index of social inequality: aggregate landholding (small to large) is used as a measure of household poverty or wealth and a predictor of household agricultural and economic productivity, food self-sufficiency, food insecurity, and technology adoption. Intolerably severe or growing inequality in landholding, experienced by peasant cultivators as denial of basic subsistence rights, has been the cause of most peasant revolutions of the twentieth century (Wolf 1969). Their struggle for land encompasses the larger struggle for individual and communal human dignity, political autonomy and identity, and a decent standard of living. The political violence and discrimination that precede conflict, the destructiveness of active conflict, and the difficulty of restoring communities and reconstructing food systems following conflict, are probably the most significant causes of food shortage and related poverty and deprivation.

The most important way in which landholding affects food supply is in the choice of crops. Plantation or latifundia systems in Central America historically have focused on cash crops for export, with smallholdings for food to meet subsistence needs. Indigenous cultivators in Guatemala historically were pushed off the most productive lands into mountainous refuge zones; in El Salvador, subsistence holdings were reduced to allow larger holdings for coffee. Struggle for land by former combatants as well as non-combatants in the former war zones of Guatemala, El Salvador, and Nicaragua remains a threat to political stability and future food self-reliance, as indigenous and peasant communities try to regroup, establish rights over land, and produce a mix of subsistence and cash crops for internal and external markets. On the island of Negros in the Philippines, large landholders employed peons to farm sugar. Landlords did not allow workers to convert land-use to subsistence production, even after a precipitous drop in the price of cane: they preferred no product to a possible loss of control to the workers. In Senegal, Franke and Chasin (1980) have documented the ways in which French control over land, and demand for the cash crop peanuts, destroyed subsistence production and the traditional symbiosis between farmers and herders.

All these cases illustrate how the total amount of food grown in any developing country, and global food production overall, are affected by land tenure. The specific impacts, in turn, of land tenure and associated crop choices on household and individual hunger depend on the relative prices of cash versus food crops, and on who benefits from cash-crop revenues. Additional factors are, on large holdings, whether workers are paid enough and food is plentiful and cheap enough to ensure a decent living; and on smaller holdings, whether the net income from cash crops more than compensates for the costs of foods that are no longer home-grown.

Crop choices and hunger impacts also depend on who controls thc income from cash crops - control that, in traditional societies, is tied to gender-based land tenure and labour exchange. In the Gambia, women refused to contribute labour to their husbands' potentially highly productive irrigated rice plots (in this case rice was a food crop sold for cash) because they would not control the income, whereas they controlled food that they grew on their own lands (Jones 1983). A 20-year perspective on Nigerian pert-urban agriculture has shown that female cultivators, drawing on kinship-based land rights, have developed an unanticipated niche that supplies greater Ibadan with roots, tubers, fruits, and vegetables (Guyer 1995). In Kenya, non-governmental organizations (NGOs) and the FAO's International Plant Genetic Resources Institute (IPGRI) have designed programmes to promote the production of indigenous food species for home consumption or market sales. Gardening everywhere is being encouraged as a way to increase food and fodder for internal markets and to improve household income and nutrition. Local gardening is also a way of conserving indigenous species in the environment.

Crop choices and production practices in much of Africa are changing in response to greater individualization of land holdings and wage labour but, as will be suggested further below, the impacts of such changes on food supply and household food security are predictably mixed, since the circumstances are so variable.

Whether the context is cash or subsistence crops, an additional supply question is whether larger or smaller holdings are more productive or efficient. Wealthier landholders are more able to invest in agriculture - to assume the risks of new agricultural technologies and the recurrent costs of seeds, fertilizers, machinery, and labour. They may enjoy also superior access to credit and economies of scale in the purchase of inputs and in control over the prices for their products. There is no question that the early adopters of GR technologies were wealthier landowners who were able to absorb the risks and enjoyed major benefits. They became wealthier also by absorbing the smallholdings of neighbours who could not compete in the new productive environment. However, it is questionable whether production per unit land generally is higher on larger holdings, because smallholders, given the means, may invest labour and other inputs more carefully and achieve even higher outputs per unit area than their wealthier neighbours. Equitable or fair access to credit, crop insurance, inputs, and markets may remove most economies of scale. But under modernizing agricultural conditions in much of the world, such conditions are hard to meet. Cooperative ownership and cultivation is one way that the possible disadvantages of smallholdings can be resolved; but, again, the evidence is mixed on whether larger communal or smaller individual holdings are more productive and efficient: so many other sociocultural and political economic factors besides landholding are involved in explaining why productivity is low on a Mexican ejido cotton cooperative or high on an Israeli citrus kibbutz.

In summary, the evidence on the ways in which land tenure influences agricultural investments and agricultural sustainability is mixed. Predictably, throughout the developing world, poverty and insufficient access to land result in unsustainable practices, including reduction of fallow, that lead to soil and water degradation, land exhaustion, and plummeting productivity. Farmers who lack resources to protect or improve land and water supplies have been blamed for soil erosion, desertification, and other ills (especially in SSA) and for deforestation and diminishing ecological resources and biodiversity in Asia and Latin America. Those working but not controlling the land can be expected to select the easiest or cheapest, rather than the most sustainable, cultivation practices. Morvaridi (1995) found that contract farmers in Cyprus used cultivation methods that were optimal only in the short run; because they did not own the land, they would not suffer the costs of long-run degradation. Conversely, extremely plentiful access to land can also discourage conservation: for example, extensive holdings in the Sudan have been tied to unsustainable agricultural methods that mine soil nutrients; after reaping initial profits, large landowners move on to exhaust additional areas. Whether owners deem it economic to maintain soil and water resources depends also on the conditions and costs of labour supply. Communally maintained terracing and waterways are falling into disrepair in countries such as the Philippines, where labour sees greater economic opportunities outside agriculture. Thus, land and labour constraints on productivity are closely tied, particularly under conditions of economic modernization and marginalization.

Labour

The amount of land that can be sown and harvested is, clearly, tied to available and affordable labour supply. Planting and harvesting are both activities that require far more labour than the rest of the agricultural cycle. In communities where these activities are shared, productivity on individual plots may be greater than if families had to provide all the labour that they could not afford to hire. Communal farming, although no longer common, provides some of the same advantages.

Where greater integration into the market economy disrupts traditional labour exchange, production may fall, as shown in Gudeman's classic study of a Panamanian village (Gudeman 1978). The reduction of patronage ties, such as in South Asian villages that have adopted GR technologies, may also produce labour bottlenecks that affect harvests. In many parts of SSA, modern cropping programmes fail where men control most of the land, technology, and proceeds but women are supposed to do much of the work, especially weeding. In Kenya, the increase in crop yield resulting from weeding was 56 per cent in female-headed households, but only 15 per cent in male-headed households; this led researchers to hypothesize that women do a less thorough job of weeding where they do not expect to control the benefits (Gittinger et al. 1990).

High labour costs may discourage extra hand cultivation and marginally lower outputs. But low agricultural wages discourage participation in the agricultural economy, where industrial or other opportunities exist. Very small household plots that cannot provide sufficient food and income push workers off the farm in search of income and can cause bottlenecks at planting and harvest times that lower food production. In Mexico, careful research into maize varieties and associated agricultural packages that would benefit small farmers proved less attractive and raised maize production less than anticipated; workers still could earn more off-farm, so abandoned farming. Scarcity of labour more than land is also a major constraint on production in much of Africa, where larger land areas since colonial times have experienced labour bottlenecks, as men were drawn off to work in the mines or to do other waged work and left women to clear, plant, and weed, with peak agricultural labour demands during the hungry season (Richards 1939). In such contexts, the problem of hunger is linked to underproduction in a vicious cycle.

Dietary Preferences

An entire class of sociocultural issues related to food shortage have to do with consumption patterns and preferences. It is easy to think of these as relatively unimportant, because a preference for maize over sorghum in a food-short region is unimportant relative to total calorie availability. However, food preferences are one set of factors that determine what foods are grown and whether farmers, as in drought-prone SSA, sow and harvest drought-tolerant sorghum or sow a riskier maize crop that totally fails in drier seasons. The World Bank has identified overconsumption of wheat and rice as a key contributory factor to the African agricultural crisis: imports of these grains have soared over the last two decades to the detriment of local production of other staple crops. Wheat and rice can both be grown in most African countries but only at costs far greater than the price of imports. Whether tastes for these imports grew in an era of economic prosperity, as was the case during the Nigerian oil boom years (Smith 1991), or as a result of dearth relieved by food aid, dependence on imports persists and has created a mismatch between economically feasible local production possibilities and consumer demand (Nwomonoh 1991). Burkina Faso, which was severely affected by the Sahelian drought of 1968-1974 and again by drought in the early 1980s, imported 94 per cent of its grain as wheat from 1966 to 1970. Wheat still constituted 52 per cent of grain imports during the first half of the 1980s, although the difference was made up by the traditional grains maize and sorghum (Barkin et al. 1990). It is not clear to what extent the continuing preference for imported wheat interferes with the country achieving food self-sufficiency during non-drought years; land, labour, access to technology, and income are probably more important factors.

POLITICAL AND ECONOMIC FACTORS

Sociocultural factors influencing tastes and land or labour allocations are closely tied to political-economic forces, especially government policies and market conditions. This section considers other causes of food shortage that are tied to governments and markets—insufficient incentives for production, trade and import constraints, structural adjustment, and food aid.

Insufficient Incentives for Food Production

Government initiatives to change consumption patterns out of economic self-interest are usually ineffective: Nigeria officially stopped rice imports in the mid-1980s in order to save foreign exchange, only to find that, by the end of the decade, rice consumption was still at two-thirds its previous level, owing to illicit re-exports of rice from Benin (Spencer et al. 1989, as cited in Reardon 1993). Efforts to protect local markets from inexpensive foreign grains (including cheaper products from neighbouring countries) are impeded by black market leakage that also distorts the food-availability situations in both donor and recipient countries. Government production and marketing-board policies, by contrast, very much influence what farmers grow. In Kenya, government crop-breeding research, extension, and grain marketing in the 1980s focused on and successfully encouraged the production of maize, not sorghum; and in Tanzania, government crop insurance encouraged farmers in drought-prone areas to assume the greater risk and grow the less drought-tolerant (but, in a good year, more profitable) maize (Louis Putterman, personal communication 1995). Overall, relative crop prices influence what is grown, sold, or later consumed and thus influence food security at the national and household levels.

Cash-Crop Versus Food-crop Promotions

Cash-crop promotions, by governments seeking to increase export earnings, may discourage production of food crops for local consumption through price, tax, and marketing structures. Some governments, as in Kenyan and Tanzanian promotions of tea and coffee, penalize farmers who wish to hedge their bets and maintain some subsistence production alongside cash crops, whose prices are outside the farmer's control. Others, as in the Mexican case cited above, provide incentives in the form of credits or crop insurance to grow particular cash crops, such as sorghum. Commercial export crops also expand relative to food crops where they enjoy the benefits of agricultural research and development, as well as priority access to critical inputs such as fertilizer (Hendry 1988). Where more resources are devoted to coffee than to millet, food availability may decline even while agriculture is modernizing, as was the case in Rwanda. Similar declines are reported where lands that could have produced basic food crops are shifted instead to more profitable livestock, forage, or animal feed (Barkin et al. 1990).

Export production can improve food availability for an individual country, if export earnings purchase more food than might otherwise have been grown. Islam (1994) demonstrated that the majority of countries that had expanded non-food production also experienced an increase in aggregate food supply. But countries that are dependent on raw-materials exports are often severely affected by even moderate shifts in world demand/prices of their products (Sheehan 1987). In addition, diversion of prime rice lands into higher-volume fruits and vegetables may account in part for the plateauing of the Asian GR in recent years.

Production incentives for crops are also limited if surpluses cannot be easily marketed. The US National Academy of Sciences (1986) noted that, in Africa, more intensive production systems tend to be located near railroad lines; in other areas, deteriorating transportation systems have discouraged remote agriculturists from responding to price increases for their products.

Price Caps on Food Crops

Rural production is, however, more constrained overall by low food prices than by inability to respond to favourable prices. A large number of countries have governmental policies capping the prices of basic foodstuffs. These cheap food policies work primarily to the advantage of urban residents but discourage investments in the agricultural sector. Price ceilings on agricultural goods depress basic food production because rural producers have less incentive to grow food crops, farm larger areas, or intensify their production methods when the rate of return for their efforts is limited. Underpricing of agricultural products on the domestic market has been linked to slow rates of adoption of new agricultural technologies in a number of countries (UNICEF 1988).

Low food prices benefit urban dwellers and manufacturers who desire to keep wages low. Urbanites typically have less access to food through non-market channels than their rural counterparts and, more importantly, they are in a better position to pressure governments into protecting their interests. Even though the majority who are poor lack political power, manufacturers who pay urban wages put pressure on governments to keep urban prices low so that they can keep costs down and compete more effectively in internal and external markets.

In addition, in many countries, peasant producers buy and sell grain on a small scale throughout the year and therefore have conflicting needs as both producers and consumers (de Alcántara 1992). This is particularly true when landholdings are not large enough to meet household food needs, given the expense of agricultural inputs. Cheap food policies may receive political support even from the rural dwellers who would benefit most directly from their elimination.

Even without price ceilings, imported grain may be cheaper than locally produced grain if it is purchased from countries such as the United States that have more efficient, larger-scale production methods and that also subsidize production. Similarly, food aid - even if given only in crisis situations - may remove local incentives to produce surplus because local producers know that emergency carryover stocks are available internationally (Brown and Kane 1994).

Local currency overvaluation also skews internal terms of trade in favour of urban dwellers since it renders imports artificially cheap, and imports drive down prices for domestic produce. The Nigerian naira was so overvalued during the oil boom that imported food was several times cheaper than locally produced food; Smith (1991) described domestic food production under these conditions as "almost irrational."

Higher prices for agriculturists in most cases increase aggregate food supply and contribute to food self-sufficiency. After years meeting its food needs by relying on lower-cost imports, the Belizean government provided incentives to agriculturists that offset the effects of controlled urban food prices; in 1981, Belize became self-sufficient in production of staple foods (Moberg 1992). However, the national marketing board operated at a loss because higher producer prices were not translated into higher consumer prices. Later, when government spending was limited by structural adjustment, the producers' price incentives were removed and the country again became a net importer of food.

Opportunity Costs and Food Production

An additional consideration is that income from food crops must also be at a high enough level to meet opportunity costs of producing food rather than other cash crops or trade or wage goods. As consumers in the developing world become more integrated into the world economy they need more cash to purchase goods that they now desire but cannot produce themselves. Migration out of rural areas - either of entire families or, more commonly, of individuals - is often a temporary or repetitive strategy to earn cash. The same factors that create one-time need for urban employment often re-create it: patterns of circular migration between cities and the countryside are often repeated year after year. In sparsely populated rural areas, especially in sub-Saharan Africa, this creates a shortage of agricultural labour and a high dependency ratio. Older people and young children often remain in rural areas while those at prime productive economic ages migrate. Households have little incentive to maximize food production if their needs are better met by participation in the cash economy, particularly if some of their needs can only be met by participation in the cash economy. Shifts to less nutritious but less labour-intensive staple crops (especially cassava) have been tied to labour shortages (Benería and Sen 1986; Bukh 1979; FAO 1987; Huffman 1987; Protein-Calorie Advisory Group of the United Nations 1977; Tabatabai 1988; Thaman and Thomas 1985; Ware and Lucas 1988).

Even households that remain oriented toward subsistence agriculture usually need off-farm employment to generate the cash necessary to purchase agricultural inputs and additional food and non-food items. Labour shortage in agriculture has also emerged as a problem in more densely populated rural areas of Latin America, where intensive labour input is necessary to sustain or raise production. De Alcántara (1992) reported that, in Mexico, emigration by some household members in order to generate income for agricultural inputs increases both the need to rely on labour-saving herbicides and the use of hired day-labour. De Alcántara also noted the consequences of labour shortage for the rural community as a whole; households with high dependency ratios have very little labour to contribute to maintaining local irrigation networks and other public works that keep agriculture productive. They also have less incentive to adopt new labour-intensive technologies based on new seeds and chemical packages, if participation in manufacturing and service industries is more remunerative. Especially where farmers do not own land but must pay rent, net revenues after expenses can reduce incentives to grow food.

Producers are also less rewarded for their efforts if a portion of their crop must be paid to the landlord, as is common in sharecropping, and especially if they must repay debts assumed over the cultivation cycle immediately after the harvest, when grain prices are usually low. Even where rents are fixed, cultivators still do not have as much economic incentive to invest in methods that can raise production by retaining soil fertility from year to year. Thus, landholding, access to credit and markets, and relative income all affect production.

Taxation Policies

State taxation policies also reduce incentives for rural agricultural production, since they transfer resources out of the rural sector. Appropriating revenue from the agricultural sector has proven to be one of the most successful short-run methods for increasing state revenue (Hinderink and Sterkenburg 1987). Production is taxed either directly or indirectly, in that many crops must be channelled through government marketing boards which, given low purchase prices, effectively appropriate a share of the produce. As the only legal consumer, marketing boards are in a position to set domestic prices that are typically held artificially low. If produce is sold at world market prices, the marketing board appropriates the difference. The effect of marketing boards is to depress producer prices, sometimes by as much as 85 per cent (Bates 1988). Another indirect tax is licensing fees: farmers must pay for the right to cultivate or sell certain crops. All of these practices decrease producers' income. A number of studies conducted in the Philippines and reviewed by UNICEF (1988) concluded that government policies to support the farm sector were insufficient to overcome the negative effects of regulated pricing, government control of trade, export taxes, export quotas, and special levies. Government control of trade in certain Philippine commodities had much the same effect as marketing boards: potential profit was shifted from farmers to government agencies (UNICEF 1988).

Insufficient incentives for food production result from the combination of government policy and market forces. Together, they have created structural problems that interfere with food production and home-generated food supplies.

Trade and Import Constraints

A country's capacity to import food to compensate for production shortfalls is limited by its export earnings. However, it is overly simplistic to think of export earnings as a direct result of domestic productivity. World terms of trade effectively discriminate against products from developing countries. Although important steps toward trade liberalization have been made in the past decade, tariffs have been reduced to a greater extent on products from developed countries than those from developing countries. Under the General Agreement on Trade and Tariffs (GATT), tariffs are levied in proportion to the level of processing, a convention that exacerbates the inequality between already industrialized nations and those in earlier phases of the industrialization process. The General System of Preferences, which provides for the non-reciprocal reduction of tariffs on developing-country exports, has done little to compensate for this structural inequality because there are multiple individually negotiated preference schemes, many loopholes, and no legal guarantees that the provisions will be followed (Toton 1982).

Within countries, both food availability and revenues from export crops are limited by the ability of producers to market their crops. One of the most positive functions of marketing boards is to facilitate connections with remote rural areas and help develop infrastructure. However, the ability of centralized efforts to improve marketing can be limited by political will, timing of efforts, and available funds. Although the Zambian government in 1985 pledged to haul all maize to safe storage before the onset of rains, the poor repair of roads - as well as the unavailability of grain bags, spare truck parts, and diesel fuel - made this impossible. Good (1986) observed that, even if the problems had been identified in time, government resources would have been insufficient to deal with them. Hence, Zambia was food short in a year with bumper harvests. In other countries, such as Belize, seasonally impassable roads are one of the reasons that urban dwellers rely on imported food (Moberg 1992).

When food is marketed between countries, transportation bottlenecks and lags can become significant issues. Some crops, particularly grains, are more durable, and less is lost if transportation is delayed at some point along the marketing route. For other crops, such as potatoes or vegetables, delays can mean seriously compromised quality or large losses.

Structural Adjustment

Many of the political and economic constraints on production are related to heavy debt burdens in developing countries, which exacerbate their marginality and powerlessness to influence terms of trade in the world economy. In addition to the direct costs of financing debt, structural adjustment programmes designed by the World Bank and the International Monetary Fund (IMF) to promote debt repayment alter production relations in ways that can increase food shortage. Structural adjustment can have positive effects on food security in the long run. Restrictions on luxury imports can free foreign exchange for more necessary imports like fertilizer. Currency devaluation, restrictions on food imports, and lifting of food price controls may even increase local production, since some of the disincentives outlined above no longer apply. Food production has increased in some structurally adjusted economies (Meller 1992; Weeks 1995).

However, structural adjustment improves security via mechanisms that require a fairly extended time-frame. In the shorter run, financial austerity programmes associated with structural adjustment more often than not restrict food availability within affected countries. In India, attempts to reduce the government deficit led to decreased expenditures on rural roads, fertilizer subsidies, irrigation, agricultural extension services, agricultural research, and other rural infrastructure (Mukherjee 1994).

Structural adjustment policies that encourage exports of both food and nonfood cash crops to meet external obligations can also create food shortage within regions. A structural adjustment process was implemented in Brazil in 1981; by 1983, per capita production of staple foods had declined by more than 15 per cent while per capita production of sugar cane increased by more than 35 per cent and cultivation of other exportables also expanded (Macedo 1988). Although basic food production rebounded somewhat in subsequent years, it did not grow as fast as export-oriented production, in spite of government programmes that attempted to expand vegetable gardening for local consumption (Macedo 1988). Similar effects have been documented in Peru in response to a more gradual adjustment programme from 1977 to 1985. Annual per capita food production fell by about 26 per cent during that period, and imports only partially compensated for the decline (Figueroa 1988). Thus, adjustment policies can have profound and long-lasting effects on the composition of agricultural production.

Food Aid

Aside from these structural or economic issues, imports are also limited by the availability of humanitarian aid. Since there is no global food shortage, food for aid is clearly "available" in an absolute sense, but political agendas of both donor and recipient countries dictate that aid must serve other than strictly humanitarian ends. This finding is no longer shocking, in that US food donations have been made selectively on the basis of military, political, or economic importance at least since the 1970s and 1980s (e.g. Wallerstein 1980), and support to countries that are not of some strategic importance remains limited, although there has been some upsurge in humanitarian emergency assistance in the mid-1990s.

Food aid has successfully expanded commercial markets for producers in donor countries; in addition to fostering heavier dependence on imports, the availability of cheap, non-traditional grains helps change consumption patterns and thereby creates future demand for these grains (Toton 1982). Potentially receiving countries may have very different strategic importance when evaluated by political or military leaders than when evaluated by agribusiness executives. A country is most likely to receive priority for food aid when there is a clear humanitarian imperative, when donor governments are sympathetic to leaders currently in power in the needy country, and when future transfers through market channels are likely. The absence of any one of these factors can be limiting.

INTERRELATIONSHIPS BETWEEN CAUSES OF SHORTAGE

The environmental, sociocultural, and political/economic causes of food shortage are hardly independent. Even though inadequate fertilizer or water may place biological limitations on production, the main causes of disruption of fertilizer supplies and irrigation systems are economic; even though drought reduces food supply, it rarely leads to food shortage in the absence of armed conflict; even though consumer tastes and preferences are socially and culturally determined, they are heavily influenced by trade and aid.

Food production generally is more and more influenced by international market and policy trends as technology extends even into remote areas and as fewer agriculturists are self-sufficient (Jazairy *et al.* 1992). The terms of market integration are especially important: smaller producers may be protected against food shortage in cases of local food-crop failure, but they may be more vulnerable to food poverty overall. New agricultural technologies have been key factors allowing food production to keep pace with population growth on the global level, but local levels of food-grain production and incomes have become more variable among those using the technology (Chattopadhyay and Spitz 1987), and affordable food must be available to compensate for shortfalls.

There are some other downsides to increased agricultural productivity. Success in market production does not necessarily mean that production areas earn or produce enough to feed their populations. Vast increases in animal feed production and produce for urban markets in Mexico were accompanied by decreased production of basic foodstuffs. The "success" of these modern agricultural programmes attracted further financial and technical assistance, which reinforced the economic rationality of neglecting basic food production (Barkin et al. 1990). Similarly, favouring of soybeans relative to black beans increased productivity and Brazilian exports, but decreased beans as affordable and essential food for Brazilian producers and consumers. Specializing in export agriculture does not doom a region to food shortage but it does increase local or statewide vulnerability to food shortage and shows that crop choices have important implications for world food supply.

THE RELATIONSHIP BETWEEN DROUGHT AND FAMINE

Some of the complex relationships between the causes of food shortage are best appreciated in the locations with prolonged drought.

They demonstrate that food shortage is not inevitable in regions that experience even major production shortfalls. There is much to be learned from cases where drought and other natural disasters did not end in famine, and particularly from developing countries that have succeeded in avoiding famine during lengthy drought.

The 1991/92 drought in Southern Africa, referred to as the "apocalypse drought" because of the magnitude of the problem, provides an unusually dramatic example of a large-scale natural disaster that resulted in very few deaths. Rains failed (or were late) across a wide region in 1991/92; the worst rainfall levels in over a century followed generally below-average rains across Southern Africa in 1989/90 and 1990/91. Grain yields in the ten states of the Southern African Development Community (SADC) were 56 per cent of normal4 (Green 1993). Regional stockpiles were woefully inadequate to cope with the shortage. The drought placed 17-20 million people at risk of starvation. Yet there were no famine-related deaths reported, except in Mozambique where there was an ongoing civil war (Callihan et al. 1994).

Famine preparedness and prompt response on the part of governments in the region to warning signs of famine are an important part of this success story. Even though regional stockpiles controlled by the SADC were insufficient to deal with the magnitude of the problem, the reserves were released onto the market early in the emergency, before food aid from other areas had arrived (Field 1995). Other interventions taken by governments in the region were far from novel but were implemented much earlier than similar strategies typically have been. Food imports and food aid, initiation or expansion of public works, and loans to agriculturists all addressed issues of supply and demand - rather than simply relief - early in the crisis (Field 1995). The government of Zimbabwe also pledged to purchase large quantities of grain before any donor aid had been committed; this proved to be a lifesaving factor (Callihan et al. 1994).

Donors can be slow. Drought is nothing new in Southern Africa. Reports of low rainfall early in the 1991/92 growing season did not raise much alarm. There was some hope that later rains would salvage some reasonable crop yields, and there was little external donor perception of an emergency, despite a fairly well developed famine early warning system. Advances in early warning technology are of little use unless the warning signals are heeded (Buchanan-Smith et al. 1994). Before any external needs-assessment teams had arrived in the region, most of the SADC National Early Warning Units had already calculated initial food needs (Callihan et al. 1994). Nevertheless, it was 4-6 months before any donor aid reached Southern Africa in 1992. Relief food would have been even slower to arrive had it not been distributed through the SADC, which collected food in distribution centres even though it had not yet been determined where it ultimately was going (Callihan et al. 1994). Good rail, road, and communications infrastructure within the SADC facilitated delivery of food from the distribution centres.

Advance procuring of grain through market channels not only helped to provide food before aid arrived but also helped to avoid the precipitous price drops often associated with sudden arrival of vast quantities of food into drought-stricken regions. Grain prices thus remained relatively stable, protecting incomes of local farmers. Food also reached needy populations before they found it necessary to leave their homes. This greatly facilitated later rehabilitation efforts, since social and production systems were not disrupted. The advance commitment on the part of Southern African governments to import grain also helped prevent prices from being driven up by speculation, as has happened in other situations where crop failure has been accompanied by insufficient confidence in the ability of the government to import food (Ravallion 1987).

In addition, most food-distribution programmes were implemented through market channels, and rural works projects prevented collapse of rural markets during the crisis (Teklu 1994). Botswana did very well with a cash-for-work relief programme that was targeted to the poor by holding wages slightly below market rates (Callihan et al. 1994). The cash-for-work programmes were part of Botswana's Inter-Ministerial Drought Committee's ongoing relief activities (Quinn et al. 1988). Botswana has already done what the SADC is encouraging all of its member nations to do: it has built the expectation of drought into its budget, instead of treating it like a shock, and it has such programmes operating and ready to expand in case of drought.

Other countries in the region made use of both food-for-work and cash-for-work programmes Almost all of the targeted food-distribution programmes were implemented through non-governmental organizations (NGOs) that had been operating in the affected communities prior to the drought. Resources came not only from the NGOs but also from proceeds from food sold through market channels. Maize subsidies were lifted in Zimbabwe and Zambia during the relief effort, to increase producer incentives at a time when large supplies of foreign maize would otherwise have driven prices down (Callihan et al. 1994). Malawi was the only country that relied on completely free distribution of food as part of its relief effort.

Although we have stressed the factors that prevented food aid from having detrimental effects on the region, the drought would almost certainly have led to famine in the absence of aid. The United States had record amounts of yellow corn on hand at the time that Southern Africa needed it most: about 12 million tons of grain were delivered in 1992 (Callihan et al. 1994). Some of the aid went through the World Food Programme and some of it was distributed through bilateral arrangements. The United States also provided US$112 million in non-food assistance, primarily in support of transportation and logistic coordination, agricultural rehabilitation and agricultural inputs, emergency water supplies, and health activities (Callihan et al. 1994). Importing was also easier during the drought, because the World Bank relaxed target dates for structural reform actions and made credit available.

Fortuitously, the bulk of grain available during the 1991/92 emergency in Southern Africa was the region's usual staple grain. Aid was then received without causing either temporary or longer-run shifts in local consumption patterns.

The lack of famine mortality, the lack of widespread social disruption, familiar relief foods, and distribution of food through market channels all made it easier for Southern Africa to recover from the disastrous agricultural conditions in 1991/92. Good weather in the following crop year (1992/93) was also critical, since it is unlikely that such a massive relief effort could have been sustained over time.

Other actions taken in response to the 1991/92 emergency made return to normal conditions quickly more plausible. The Sorghum and Millet Improvement Program of the International Crops Research Institute for the Semi-Arid Tropics (ICRISAT) in Matopos, Zimbabwe, provided (with funds from USAID) improved and tested varieties of drought-resistant small grains that matured earlier than the traditional varieties. These were approved for use in all of the SADC countries except Malawi - but even Malawi had a record agricultural harvest in 1993, in part due to improved maize seed that was distributed by an NGO (Callihan et al. 1994). Programmes had also expanded within the SADC to preserve cattle during times of drought, in order to help protect future livelihoods. These types of interventions were possible only because people in the region were not pushed into famine conditions under which they would have chosen short-run survival strategies over long-run subsistence strategies (Field 1995).

The experience of Southern Africa during the 1991/92 drought is not a complete success story. Mozambique fared less well than other Southern African countries, in part because donors were reluctant to send food aid that could be stolen by the Mozambican armed forces and not reach displaced people (Ayisi 1992). Production also did not rebound with the good rains in 1992/93 to the same extent as it did in neighbouring nations, largely owing to the ongoing civil war. Another, less severe, drought afflicted the region in 1994/95, and farmers who might otherwise have been able to cope were pushed into bankruptcy, since they were already in debt from the 1991/92 drought. Nevertheless, the experience during the worst drought in over a century clearly shows that drought does not have to lead to famine. The physical and biological causes of production shortfalls are in no way the sole determinants of food shortage. They must always be viewed against an institutional background dedicated to preventing and alleviating shortage.

ECOLOGICAL AND POLITICAL ASPECTS OF FOOD SHORTAGE IN THE 1990S

In evaluating the causes of food shortage, politics has been implicated more than the weather. This is because politicians shape the environment of response to ecological conditions. They also shape the trade-and-aid policies that determine whether households, regions, and countries produce enough food to provision themselves or have affordable terms for import and purchase. National politicians and policies also determine the extent to which regions and localities can retain or develop food self-reliance. Throughout much of the developing world, small farmers have capacities to improve production but lack certain access to land, moisture, seeds, and markets to make optimal use of that potential. They also lack access to basic services, such as health and education, that could improve their lives and prevent food shortage.

Since the 1990s, the international (UN) community has sponsored a number of century-end summit meetings to take stock of current resources and to plan for the future: these were the (UNICEF) World Summit for Children (1990), UNICEF-WHO Conference on Ending Hidden Hunger (1991), International Conference on Nutrition (ICN, 1992), UN Conference on Environment and Development (UNCED, 1992), World Conference on Human Rights (1993), International Conference on Population and Development (1994), World Summit on Social Development (1994), Fourth World Conference on Women (1995), and World Food Summit (WFS, 1996). Almost all addressed two principal dimensions of food shortage and its prevention - the need for better mapping of hunger vulnerability, and the need for more stable political environments for food security. Although especially UNCED, the ICN, and the WFS addressed a plethora of additional technical, economic, social, and cultural issues surrounding food security now and into the twenty-first century, these two dimensions are probably the most significant for addressing local to global food-shortage problems. Implementation of the first is likely to be assisted by the momentum of political will generated by the various summits; implementation of the second is unlikely to be affected by international proclamations. In philosophical or humanitarian terms, few disagree that adequate food is a human right. But most continue to disagree over how to achieve universal food security in a world context divided by socio-economic inequalities, ethnic differences, and narrower country-level political interests. Biotechnological initiative that may break the "yield barrier," and adaptive research and extension to narrow the "yield gap" in basic foods, may help production keep up with population growth and prevent food shortage. But achieving food security for households and individuals remains a greater challenge.

6

Agricultural Development in Developing and Developed Countries

INTRODUCTION

Agriculture is a sector that plays a central role in terms of production, employment, foreign exchange generation, etc. in developing countries. Agricultural trade has however been lagging significantly behind trade in manufactured products. World trade in all manufactured products expanded at 5.8 percent from 1985 to 1994 whereas agricultural trade grew at only 1.8 percent during the same period (Binswanger and Lutz; 2000). Moreover, Africa is the hardest hit by the unfair world trade. The share of total developing country exports in world exports increased from 19 percent in 1973 to 28 percent in 1980 (partly due to high oil prices) and remained stable at 22 to 23 percent thereafter, that share was 40 percent in 1961. From 1985 to 1995 the Asia share increased from 10 percent to 15 percent whereas the African dropped from about 4 percent to about 2 percent (WTO; 1996). The Middle Eastern countries also lost about half of their market share, while Latin America largely held its ground.

International trade rules can create an enabling environment for poverty reduction. Good international rules can facilitate policies that are good for the poor. However, many of the rules provided in the WTO failed to be good, threatening to marginalize developing countries and the world's poorest people within an already unequal global trading system.

Economic development of developing countries is hampered by agriculture. Agricultural development itself is hampered by what has been happening in the world trade. Therefore, the purpose of this paper is to address issues including : (1) what has been the relationship between industrialized and developing countries as regards to trade in agriculture; (2) theory, practice and consequences of supranational organizations; (3) what loopholes in rules and tricks used in world trade to block the export of agricultural commodities of developing countries signifying the unfair world trade system; and (4) what implications globalization process has on Africa.

THE RELATIONSHIP BETWEEN INDUSTRIALIZED COUNTRIES AND DEVELOPING COUNTRIES AS REGARDS TO TRADE IN AGRICULTURAL COMMODITIES

The full potential of trade to reduce poverty cannot be realized unless poor countries have access to markets in rich countries. Unfortunately, Northern governments reserve their most restrictive trade barriers for the world's poorest people.

When desperately poor small holder farmers or women garment workers enter world markets, they face import barriers four times as high as those faced by producers in rich countries. Trade restrictions in rich countries cost developing countries around $100 bn a year—twice as much as they receive in aid. Sub-Saharan Africa, the world's poorest region, loses some $2 bn a year, India and China in excess of $3 bn. These are only the immediate costs. The long-term costs associated with lost opportunities for investment and the loss of economic dynamism are much greater.

Trade barriers in rich countries are especially damaging to the poor, because they produce, such as labour-intensive agricultural and manufactured products. Because women account for a large share of employment in labour-intensive export industries, they bear a disproportionate share in industries, they bear a disproportionate share of the burden associated with the lower wages and restricted employment opportunities imposed by protectionism.

Nowhere are the double standards of industrialized country governments more apparent than in agriculture. Total subsidies to domestic farmers in these countries amount to more than $1 bn a day. These subsidies, the benefits of which accrue almost entirely to the wealthiest farmers, cause massive environmental damage. They also generate over-production. The resulting surpluses are dumped on world markets with the help of yet more subsidies, financed by taxpayers and consumers. Oxfam has developed new measures of the scale of export dumping by the European Union (EU) and the United States. It suggests that both these agricultural superpowers are exporting at price more than one-third lower than the costs of production. These subsidized exports from rich countries are driving down prices for exports from developing countries, and devastating the prospects for smallholder agriculture. In countries such as Haiti, Mexico, and Jamaica, heavily subsidized imports of cheap food are destroying local markets.

Rich countries have systematically ranged on their commitments to improve market access for poor countries. Instead of reducing their own farm subsidies, they have increased them. They have liberalized fewer than one-quarter of the products for which they had agreed to open their markets.

The developing countries on the other hand are forced to open their economies by rich countries. However, rapid import liberalization in developing countries has often intensified poverty and inequality. Loan conditions attached to IMF and World Bank programmes are a major part of the problem. The IMF, the World Bank, advocacy has been backed by loan conditions which require countries to reduce their trade barriers. Partly as a result of these loan conditions, poor countries have been opening up their economies much more rapidly than rich countries. Average import tariffs have been halved in sub-Saharan Africa and South Asia and cut by two-thirds in Latin America and East Asia.

THE EVOLUTION AND DEVELOPMENT OF INTERNATIONAL TRADE AGREEMENTS

The evolution of international trade agreement dates back to the period immediately after the end of the Second World War. The proposal of establishing an international trade organization was first tabled at the Bretton Woods conference (named after the meeting place, Bretton Wood, New Hapshire, U.S.A.) in 1946, when the victorious countries (especially Britain and the United States) started to plan for a new international economic order. Accordingly, the world economy was to be organized around three cornerstones: the International Monetary Fund (IMF), the International Trade Organization (ITO), and the International Bank for Reconstruction and Development (commonly known as the World Bank). The ITO never got off the ground, partly because of disagreement between Britain and USA over the extent of the authority of the proposed ITO over the actions of governments (Sodersten and Reed 1994).

In place of the ITO, the General Agreements on Trade and Tariff (GATT) emerged as the framework for trade relations among nations. GATT was aimed at reversing the move toward protectionism in the 1930s and preventing such a move being repeated.

GATT has been functioning on the basis of the principles related to non-discrimination (which accepted the so-called most-favoured-nation clause (MFN); reciprocity (which implies a country which receives concession from another country should offer an equivalent concession in return); and transparency (which refers to forbidding of direct control on trade through quantitative restrictions except under a few defined circumstances such as balance of payments crisis). Systems of preferences are also other issues in international trade including tariff preferences; the common-wealth preference systems; and the Lome Convention i.e. covering trade between EU and Africa, Caribbean, and Pacific countries. Moreover dumping i.e. selling goods below "normal value" capable of destroying domestic industry and multi-fiber arrangement are also exceptions representing major departures from the principles of GATT. Textiles were not part of GATT negotiations. In addition, there are two even more important sectors, which are outside the jurisdiction of GATT: Service and agriculture. (Degene; 2000)

One other international institution related to trade and development is the United Nations Conference on Trade and Development (UNCTAD). It made its first conference in 1964 in response to pressure exerted at the UN general assembly for a special trade conference within the Economic and Social Council. Since its establishment 1964, a succession of conferences has been held in different places.

The organization of LDCs within UNCTAD is known as the "Group of 77" (the group now numbers 128 countries). Despite wide political, ideological and cultural differences, this group has had some success in coordinating the attitudes of developing countries within UNCTAD and so has contributed toward the adoption of a common negotiation stance in other UN bodies. The "success" of UNCTAD has been witnessed by the establishment of the Generalized System of Tariff Preferences (GSP), and International Commodity Agreements and Compensation Finance Schemes (Dejene, 2000).

UNCTAD has also helped the establishment of international commodity agreements directed at stabilizing commodity prices and compensatory finance schemes designed to stabilize export earnings of developing countries.

Eight rounds of negotiations have been under taken under the auspices of GATT. The GATT started with a round of negotiations in 1947 including: the Genera Round (1947), the Kennedy Round (1964-67), the Tokyo Round (1973-97), and the Uruguay Round (1986). Agricultural products were however, generally excluded from all of the rounds of negotiations.

The Eighth round i.e., the Uruguay Round trade negotiations, gave birth to the World Trade Organization (WTO) in 1994. WTO is established with vastly expanded scope of the multilateral trade system so that it no larger deals only with the conduct of trade in manufactures (as did the old GATT). Its scope expanded to cover trade in agriculture, trade and investment in services, and beyond trade issues, covers intellectual property rights and investment-related measures. Moreover it directed that the new issues of trade and environment be discussed at committee level in the WTO.

PRINCIPLES/RULES OR OBJECTIVES OF SUPRANATIONAL TRADE ORGANIZATION'S

THE GENERAL AGREEMENTS ON TRADE AND TARIFF (GATT)

The International Trade Organization (ITO) of 1946 never got off the ground, partly because of disagreement between Britain and USA (Sodersten and Reed; 1994). It was replaced by GATT. The objectives of GATT were: (1) to provide a framework for the conduct of trade relations among nations; (2) to promote progressive elimination of trade barriers; and (3) to provide a set of rules (code of conduct) that would inhibit countries from taking unilateral action. The first two objectives have been more or less successful, while the third one failed to materialize. The major powers were unwilling to abandon unilateral measures consistent with their national interests. For example the United States has taken several unilateral actions against countries that, it alleges, are trading "unfairly".

GATT is based on the principles of non-discrimination, reciprocity, and transparency. Non-discrimination means that contracting parties (signatories) accept the so-called most-avoured-nation clause (MFN). That is, a country agrees not to give better treatment to any single nation than it gives to all the other contracting parties of GATT. This clause (contained in Article 1) rules out any preferential treatment among nations as far as trade policy is concerned. The MFN clause does not apply to certain cases like free trade areas, customs union and other regional groupings, preferences offered to developing countries and to subgroups of those countries and trade textiles and clothing. The MFN clause has played an important role in encouraging countries to negotiate on trade liberalization.

Reciprocity has never been formally defined, but has nevertheless been cornerstone of GATT negotiations. This obligation requires that a country receiving a concession from another should offer an equivalent concession in return. Transparency on the other hand forbids the use of direct controls on trade, particularly quantitative restrictions, except under a few defined circumstances such as balance of payments crisis which the argument allowed member countries to apply quantitative import restrictions in order to deal with severe current account deficit problems. Tariff provisions for specified groups of countries; systems of preferences which were in operation before the signing of the GATT were allowed to continue; Lome convention; trade between EU; and African, Caribbean, and pacific countries; dumping or selling goods below "normal value" causing "material injury" or "materially retarding" the establishment of a domestic industry; and the multi-fiber arrangement, are cases of a major departure from the principles of GATT. Textiles, services and agriculture are not within the jurisdiction of GATT.

THE UNITED NATIONS CONFERENCE ON TRADE AND DEVELOPMENT (UNCTAD)

The success of UNCTAD of 1964, has been witnessed by the establishment of the Generalized system of Tariff Preferences (GSP) and International Commodity Agreements and Compensatory Finance Schemes.

The Generalized System of Tariff preferences was set up in response to the demand of developing countries for industrialization through export of manufactured goods. The proposal for a GSP, which emerged from the first UNCTAD conference in 1964, was based on the principle that developing countries required preferential tariff treatment without reciprocation on their part. The GSP system was negotiated at length over the period 1964 -1971. The tariff treatment of GSP did not extend to agricultural and fishery products. Textile products were excluded outright by the United States and Japan. The net benefits of GSP were limited. One reason for this is that the coverage of the GSP was limited to less than 20 percent of all developing countries' exports. In addition, some 80 percent of the imports under the GSP came from seven newly industrialized countries (i.e., Brazil, Hong Kong, Israel, Korea, Mexico, Singapore, and Taiwan). The GSP scheme (around 1980) provided a limited benefit to many of the more advanced developing countries and no benefit to others. Further, many of the goods that are of most interest to developing countries (e.g. Agriculture, textile) were excluded from GSP (Degene; 2000).

THE URUGUAY ROUND (UR)

The Uruguay round is the eighth round of negotiation undertaken under the auspices of GATT. In the first round, the

Geneva round, twenty-three countries took part. One of the interesting GATT rounds, the Kennedy Round (1964-67), was initiated by the Kennedy administration to protect American interests in the face of two emerging economic blocks (EU and Japan). The Kennedy Round tariff reductions by developed countries have been estimated at about 36 to 39 percent, and affected around 75 percent (by value) of world trade. They were however "concentrated upon manufactured goods that were of principal interest to developed countries or on the raw materials that were essential to their industry" (Sodersten & Reed 1994: 363). Agricultural products were generally excluded from the Kennedy Round because of the conflict of interest between USA and Europe regarding export subsidies (of EU) and production subsidies (of USA).

The two protagonists also disagreed on the formula for cutting tariff. Similarly, tropical products were excluded from the Kennedy Round negotiations.

The Tokyo Round (1973-97) achieved further liberalization of trade through tariff reductions. However, it did not resolve agricultural problems. Developing countries remained dissatisfied with their failure to achieve greater concessions, although some of their major products (textiles and clothing, leather, footwear) received lower tariff cuts and tropical products gained some concessions.

The Uruguay Round was basically initiated by the USA, who lost comparative advantage in several of its traditional exporting sectors (Such as steel and the automobile industry). These sectors already enjoyed considerable protection from imports, and had powerful domestic lobby groups defending their interests. The US sought to introduce to the agenda for the proposed GATT talks several new items to further protect its disadvantaged traditional exporting sectors (Sodersten & Reed 1994). These new issues included trade-related investment measures (TRIMs), such as local content and minimum export requirements, trade aspects of international property rights (TRIPs), and trade in services. The argument that countries should not expect to enjoy the benefit of free trade of they themselves do not practice fair trade has been a recurrent theme in American trade policy.

The Uruguay Round negotiation was launched on September 20, 1986 at the South American seaside summer holiday resort town of punta del Este (Uruguay). This multilateral trade negotiation was the eighth of series of negotiations undertaken under the auspices of GATT. It has been observed, "the Third World Countries were virtually dragged into the negotiations, much against their will and better judgement". The motive behind the Uruguay Round was to protect the economies of industrialized countries and to prevent emergence of competition elsewhere (Raghavan; 1990: 32-38).

The launching of the Uruguay Round negotiations and subsequent developments in global agreements have been associated with the relative decline of the US economy and the emergence of strong economies elsewhere (e.g. Japan, Europe, south East Asia) Raghavan; 1990).

The Uruguay Round negotiations were facilitated by partial break up of Third World Unity, which was fuelled by the USA and other industrialized countries. It was observed that by the end of 1985 and early 1986, Third World Unity faded away, under bilateral pressures applied by the US [Raghavan 199 : 76]. For example, the Cairns group, which represented interests of a group of countries primarily depend on agriculture for external trade, was alienated from many of the Third World Countries. A number of countries, under the leadership of Colombia and Uruguay, abandoned their opposition and/or provided some support to the US on the new round or remained silent.

However, a small group of countries, led by Brazil and India, remained together and managed to moreover within the limited space available to them, and exploited the US-EU differences to forge a tactical alliance with the latter (EU). Double standards in the theory and practices of free trade is a common feature of the Uruguay Round negotiations. The industrialized countries have often failed to practice what they preach. Most often they abandon the comparative advantage argument and protected their own domestic industries and exercised monopolies power over intellectual property. The transnational corporations (TNCs) are a major source of technologies, and they control these technologies through patents, investments, cross-licensing and services (Degene; 2000).

THE WORLD TRADE ORGANIZATION (WTO)

The Uruguay Round trade negotiations, which gave birth to the World Trade Organization in 1994, vastly expanded the scope of the multilateral trade system so that it no longer deals only with the conduct of trade in manufactures. Its scope expanded to cover trade in agriculture, trade and investment in services, and beyond trade issues, covers intellectual property rights and investment-related measures. Moreover it directed that the new issues of trade and environment be discussed at committee level in the WTO.

The WTO is based on the following principles: (1) there should be non discrimination (the MFN treatment); (2) free trade should be encouraged gradually through negotiation; (3) an agreement not to raise trade barriers can be as important as an agreement to lower barriers; and (4) competition should be encouraged.

In a sense, the UR complements what structural adjustment programmes (SAP) are achieving. The Round will lead to a very significant external liberalization of many sections and facts of the domestic economy of all the developing country members of the WTO.

Domestic laws and policies in a wide range of areas have to be changed to bring them in line with global trade rules. According to several analyses, the UR agreements will severally restrict or constrain the possible policy options in many areas. Non-compliance with the rules can result in complaints being brought against a country and the threat of trade penalties and retaliation through measures affecting trade and other activities.

Thus, signing a WTO agreement is a very serious undertaking. It is different from signing a UN Declaration, even a UN declaration of over a hundred heads of government, has little enforcement possibility and becomes only a moral commitment.

The disciplines of the WTO are legally binding on present and future governments. Once the WTO agreements come into force, it would be difficult for a government to have economic policies relating to foreign trade, investment, sectoral policies in services and agriculture, or technology policy.

The WTO is also more than a mere agreement. It has a fixed institution with a permanent Dispute Settlement Body (DSB). Decisions by the DSB, or one of the adhoc panels set up to deal with a particular dispute and composed variously of three trade experts, no longer require (as did the old GATT) the agreement of the defeated party. Compared with GATT, the WTO also has much more power and justification, but apparently it is not more open to the public and democratic control of it is no stronger than was the case with its forerunner.

IMPLEMENTATION AND CONSEQUENCES OF RULES OF INTERNATIONAL TRADE IN AGRICULTURAL COMMODITIES

The international trade, rules, principles, agreements and their implementation have very little impacts on agricultural commodities. Until the formation of World Trade Organization (WTO) in 1994 trade agreements excluded agriculture. Agriculture was outside the trade agreements undertaken between 1946 (the year in which ITO was organized) and 1994 (in which the WTO was established). For example, as ITO could not get off the ground, it was replaced by the GATT. The rules and principles of GATT were not applicable to agricultural products for agriculture together with services, and textiles were not part of negotiations. Even in the case of UNCTAD, the Generalized System of Tariff Preferences (GSP) and International Commodity Agreements and Compensatory Finance Schemes, excluded agriculture and textiles from the treatment. The Uruguay Round negotiation of 1986, similarly dealt only with the conduct of trade in manufactures. It was with the formation of WTO of 1994 that the scope of the multilateral system has vastly expanded to cover trade in agriculture, and trade and investment in services. (COMESA; 1999 Gueye and et al; 1998)

The North's motives for introducing trade-related intellectual property rights (TRIPS) (Gueye et al; 1998) in the round were to enable their firms to capture more profits through higher prices, and through royalties and the sale of technology products; and to place stiff barriers preventing the technological development of potential new rivals from the south.

The device of linking trade with other issues such as "trade and environment", "trade and labour standards" and "trade and investment" is being increasingly used for the purpose of further opening up developing economies (Oxfam; 2002) or to reduce their competitiveness in the scramble for world market shares.

There is also the intention to set the developing countries to join a new WTO agreement whereby decisions on government procurement (or government spending on supplies or projects) should be subject to WTO principles such as national treatment. In other words, it should be illegal for a government to give preference to local companies or citizens when they buy stationery or give a contract for building roads and bridges. Foreign firms must be given the same rights as (or more rights than) local firms in bidding for government business.

Concerning agricultural exports, developed countries (especially US and EU) reached agreement without direct participation of developing countries (Blair House agreement). Essentially, the agreement covers the following points: conversion of non-tariff import restrictions to fixed tariffs; reduction of tariffs by an average of 37 percent for tropical products (although the conversion of the other import barriers to tariffs does not really permit such a computation); and reduction of export subsidies and national support measures for farmers (Dejene; 2000).

As a result of the Blair House agreement, the world market price for cereals rose 50 percent to 75 percent between the summers of 1994 and 1996, while food aid declined markedly. The developing countries cut back their food imports substantially.

The WTO is still an organization composed exclusively of government representatives who take important decisions in small groups behind closed doors. Transparency and possibilities for participation of the LDCs are in fact not in place, unless they organize themselves behind spokesman for developing countries such as India or Malaysia.

The WTO suffered a severe blow at the Seattle conference of Ministers of Trade, which was held in November 1999. This conference was beset, right from the beginning, not only by a huge anti-WTO demonstration, but it was also disband without reaching any agreement. Perhaps, the outcome of the Seattle conference signals the beginning of the reversal of the globalization process (Dejene; 2000).

Contentious issues that led to failure of the Seattle conference included the following: market access concerns (particularly for textiles and agricultural products), agricultural subsidies, developing countries trade liberalization obligations, infant industry protection, trade restriction undertaken for balance of payments reasons, technical barriers to trade, and problems encountered in the implementation of TRIPs, etc.

One of the contentious issues raised at the Seattle conference was that of agricultural subsidies. Developing countries demanded the reduction or elimination of government subsidies for agricultural production and exports (COMESA; 1999). They argued that the heavily protected market of the EU, Norway, Japan, etc., should be opened to their exports. Here, we should note that most African countries do not subsidize their exports. Rather they argued for the right to subsidize their own agricultural production activities to address their non-trade concerns such as food security.

One of the most important achievements of Uruguay Round Agreements (URA) was to bring agriculture under normal WTO disciplines because the original GATT excluded much of agricultural trade from liberalization. The agreement on agriculture covers three major areas in which members have made commitments namely: market accesses, domestic support to producers and export subsidies (COMESA; 1999).

The commitments of members including both developed and developing countries are as summarized from COMESA (1999) given below:

(1) Market Access 'Tariffs Only', Which Covers Four Major Areas. i.e.,

(a) Tariffication of non-tariff barriers (NTBs)-quantitative restrictions, import licensing, variable import levies, minimum import pricing, voluntary export restrictions etc which are considered to be the barriers to market access to agricultural trade, where members agreed to convert NTBs into tariff equivalents (TE) to achieve uniformity in reduction of these barriers; in which; TE = [(internal price-external price)/ external price].

Whereby-Internal prices are representative of wholesale prices,

- External prices are the actual c.i.f. unit import price; and
- 1986-1988 is the base year for average annual data.

  Such measure, however, resulted in very high tariffs i.e. dirty tariffication; less commitment in agriculture and more use of the countries right to apply domestic support measures and export subsidies which however, they did not have or cannot afford.

(b) Tariff reduction. Tariffs have to be reduced by a simple average of 36% for developed countries over the period 1995-2000 (with a minimum of 15%) while developing countries committed themselves to an average reduction of 24% (with a minimum of 10%) over the same period.

Tariff reductions have lead to the following outcomes: average agricultural tariffs remained higher than industrial tariffs because tariffication resulted in higher tariff protection i.e., dirty tariffication, etc.; increased tariff dispersion, i.e., uneven tariff acts across different products allowing countries to retain prohibitively high tariffs on sensitive products; and the structure of agricultural tariffs has become complete with frequent use of specific and other non ad valorem (NAV) rates.

(c) Special Safeguard (SSG) measures may be taken against imports of designated terrified products. These measures constitute an additional duty on imports at a calculated rate. The problem is that since SSG can be applied to terrified products, applying them grants double protection.

(d) To ensure a swift transition to a tariff only regime at guaranteed imports level, members agreed on market access opportunities through the use of tariff rate quotas (TRQ). First, they agreed to maintain current market access i.e., the import level at least equal to average of annual imports in the base year (1986-88) by prescribing very low tariffs on those imports. Secondly, they to provide minimum access opportunities by very low tariffs for import quantities not less than 3% of average imports in 1995 and being raised to 5% of average imports by 2000. Finally, for countries not wishing to tariff their NTBs, they agreed to provide special minimum market access opportunities for imports by providing import opportunity in 1995 to the extent of 4% of the average base year and 0.85 of the base year there after.

(e) Market access problems for COMESA countries implementation of the current market opportunities through TRQs has revealed some problems. Not all import quantities under TRQs can be field due to problems related to administrative methods, lack of demand for imports etc.

(2) Domestic Support to Producers

A number of governments provide support to domestic agricultural producers in the form of subsidies, which are inherently trade distortive and undermine the basic principles of WTO. To address this problem, commitments under domestic support cover 3 major areas namely; reduction of trade distorting domestic support measures, exemptions from reduction commitments and the peace clause.

(a) Reductions of domestic support members agreed to reduce the domestic support by 20 percent. However, the reduction commitments have led to some imbalanced distribution among countries. Countries which have high support (developed countries) are allowed to maintain them with gradual reduction while those without such measures (developing countries) are not allowed to introduce them. Some developed countries still have high levels of support even after the 20 percent reduction.

(b) Exemptions from reduction commitments on domestic support including: investment subsidies to agriculture in developing countries; subsidies on agricultural inputs such as mechanization of agriculture, development of land, seed, fertilizer irrigation, pesticide etc in LDCs; support to diversification; general services like research, pest, and disease control, training, marketing and promotion of infrastructure services; public stock holding for food security reasons; domestic food aid; etc.

(c) Peace clause in the above list of items of exemption in (b) are fully exempted from any action against subsidies set out in the subsidies agreement. The peace clause was to expire in the year 2003.

(3) Export Subsidy Reduction Commitments

Export subsidies in agriculture are an exception to WTO provisions that generally are against export subsidies. So far the export subsidies in agriculture are concentrated in a small group of developed countries. According to a study by WTO, in 1995 subsidies provided by 6 industrial economies accounted for 75 percent of global value of total reductions commitments while the share of all developing countries combined was just over 20 percent. Commitments of export subsidy have five major parts: reduction commitments (that members agreed to reduce export subsidies by 36% and 24% for developed and developing countries respectively); implementation procedures; circumvention of commitments, i.e., prevents the members from creating new export subsidies, treatment of export quantities as food aid.

International trade rules matter. They can create an enabling environment for poverty reduction, or a disabling one. Good international rules can facilitate policies that are good for the poor, conversely, bad rules can outlaw policies. Many of the rules given in WTO fall into the latter category (Oxfam; 2002). They threaten to marginalize developing countries and the world's poorest people within an already unequal global trading system.

The problem is that many WTO agreements, and the manner of their implementation, reflect the negotiating strength of Northern governments, and the influence of powerful transnational companies. Lacking the economic power and the retaliatory capacity to pursue developing countries need multilateralism to work. But for multilateralism to work, it has to be fair and balanced. It has to protect weak countries from the abuse of economic power, rather than concentrate advantage in the hands of rich countries. The WTO fails the test in many areas (Oxfam; 2002).

LOOPHOLES IN THE RULES OF WORLD TRADE AND TRICKS THAT BLOCKED THE AGRICULTURAL EXPORTS OF DEVELOPING COUNTRIES

The discrimination in trade policy which affect the developing countries' exports are found in three areas:

(1) The generally high level of industrialized nations' import tariff;

(2) the escalation of the tariffs which increases with each processing stage; and

(3) the great number of non-tariff barriers to trade which impede imports, particularly of processed products. What has happened in recent years with regard to these three areas in a way depict the failure of WTO to be up to its expectations in making fair the international trade in agricultural commodities.

Tariffs

In the case of tariffs, the inclusion of the agricultural sector within the general principles of liberalization and non-discrimination of foreign supplies in WTO since 1994 has in fact reduced some distortions on the world agricultural produce markets, but the situation is still far from being a qualitative break through. The average tariff on agricultural products following the Uruguay Round is 40 percent, while on industrial products it is slightly less than 4 percent. This comparison alone makes clear the extent to which the problem still exists.

WTO members committed themselves to convert the non-tariff barriers to trade (NTB) into tariffs whose level is to correspond about the same as that of the previous NTB protection (tariff equivalents). For products for which there were no NTB, the tariff rate during the Uruguay Round's reference period of 1986-88 is to be taken as the basic tariff. Besides the commitment to convert the NTB into tariff equivalents, the industrialized nations also undertook to reduce the equivalents and the basic tariffs by 36 percent until 2004, although LDCs were exempted from this obligation.

But in reality there has been little success in putting these provisions into practice. In converting their NTB into tariffs, the industrialized nations have often fixed them at a high level. Exports estimate that "dirty tariffication" was involved in 60 percent of all EU tariff equivalents, and in 45 percent of the USA's. That means the tariffs were set so high that even after the promised 36 percent reduction the effective protection is still as great as before or even greater. In addition, since the basic tariffs were very high due to the selection of the reference period of 1986-88 this favoured the trend to higher tariffs in the agricultural sector.

In general, many tricks and a great lack of transparency can be noted in the implementation of the Agreement on Agriculture, meaning that the total burden of tariffs was scarcely reduced in substantial terms. The commitment to cut tariffs by 36 percent applies to the average tariff on all products; in the case of some individual products a reduction of only 15 percent is permitted. That is why in the case of highly-sensitive products, such as wheat, sugar, beef, maize and barley, the industrialized nations often have reduced the tariff rate by the minimum of only 15 percent. At the same time, in the case of products with an anyway low tariff rate of 3-4 percent they have frequently cut tariffs by 100 percent, which helps them to show total reduction of 36 percent when working out the average.

Experience with quotas to date makes clear they should be used for products for which otherwise there are such high tariffs that importing them is impossible. At first sight that appears to be helpful. But many quotas are 'under filled' because of tricks that hamper this. For instance, licenses are granted to suppliers that are unable to fill their entire quota.

In general, according to a World Bank study, following the Uruguay Round tariffs reduced just about one-quarter of all tariffs at most. In the case of 'tariffied' products the reduction rate is in fact only 14 percent.

Tariff Escalation

The instrument of tariff escalation, by which tariffs rise in line with processing stages, is particularly detrimental to the goal of building up processing industries in developing countries.

The latest WTO overview shows, for example, that within the EU raw cocoa has a tariff of zero percent. At its first processing stage (cocoa butter) it is charged 9 percent, and at its second stage (cocoa paste) it attracts 21 percent. The figures for coffee are 4 percent for the raw product and 11 percent for its second processing stage, and for soy beans zero percent and 6 percent respectively. Japan and the USA apply comparable scales.

A study also indicated that the proportion of processed products to the LDC's total agricultural produce exports dropped from 27 percent to 16.9 percent from 1964 to 1994, while that of the developing countries as a whole during the same period increased from 41.7 percent to 54.1 percent. This, however, covers mostly only first-stage processing. If a further processing stage is taken into account, the proportions are much lower at 8.4 percent and 16.6 percent respectively.

Therefore it is not surprising that the world's biggest exporting nations are the industrialized states. They process the developing countries' raw materials and export them as processed products. Among the five leading exporters of agricultural products in terms of value besides the USA, France, Canada and the Netherlands is Germany, which has a world market share of 5 percent, while Brazil, the giant of the agricultural exporting developing countries, is at 12th place with a 2.8 percent global market share.

Non-Tariff Barriers to Trade

With regard to NTB, there has been progress to some extent due to the Uruguay Round's constraints on tariff setting, when many previous NTB such as variable import levies, import quotas and tariff rate quotas were converted to tariffs, but with the limits described above. Many other foreign trade regulations also have the effect of NTB, and their significance has not been reduced comparably. The lack of transparency in the tariff structure and the tariff quota distribution examined above are as similar to NTB as the many health and hygiene standards, which are particularly important in the case of agricultural products. However, it should be noted that as a rule these standards are laid down to protect consumers in importing countries. They can, of course, deliberately or unintentionally, also have protectionist impacts or at least make it difficult for developing country exporters to supply markets in Europe or Japan if they demand expensive inspections, and so on.

In this area, there are two special WTO agreements, i.e., the agreement on sanitary and phytosanitary measures (SPS) and the Agreement on Technical Barriers to Trade (TBT).

Low Benefits of Preferential Agreements

Preferential tariff agreements have long been used to improve market access for developing countries by giving them favoured treatment. It general, however, it can be noted that the preferential agreements have had little success in boosting the developing countries' exports. The African, Caribbean, and Pacific (ACP) countries have not been able to increase their share of EU imports over the last 20 years. They have also been unsuccessful in diversifying their product range to any appreciable degree. Too many exceptions, especially in the agricultural sector which is so central for developing countries, strict rules on the origin of products, the exclusion of certain-sensitive products, and preference erosion have has a negative impact on agreements.

Another important factor is that favourable trade rules alone do not guarantee that the poorest countries can use their market opportunities. In those countries, industries must also be built up that can process raw materials and export finished products. The countries must promote such industries, such as by tax relief or assistance with their infrastructure. But pointing out these homemade solutions is not to diminish the importance of the preferential agreement, especially for the poorer developing countries.

The Market Power of Agricultural Exporters

In the case of most agricultural products only three to six companies are now involved in their worldwide trade. The companies' importance is in fact increasing because many developing countries have privatized the state enterprises that earlier handled their export business in the agricultural sector, which often mean selling them to transnational corporations. This was due to the pressure put on these countries by political recommendations in the context of structural adjustment programmes or in connection with the WTO debate. Therefore, for developing countries, agricultural trade is increasingly also becoming a question of access to the trade and sales opportunities of the transnational companies.

The general conditions of the world economy still have many disadvantages for poorer developing countries in particular. True, that the developing countries are also responsible for their economic plight because they have failed to improve their domestic framework conditions, but this should not lead one to underestimate the significance of the framework conditions of the world economy.

The self-interests of the industrialized nations in trade policy still hinder further reform steps. Whereas many developing countries in recent years have opened their markets, including in the agricultural sector, under the pressure of the WTO negotiations and structural adjustment programmes, the North still keep its markets closed and distorts the market even more by huge subsidies to promote its own agricultural sectors.

It is true that the WTO Agriculture Agreement also contains the commitment to reduce national agricultural subsides by 20 percent, but that leaves an 80 percent opportunity for support. Due to the Uruguay Round the developing countries suffered losses of their agricultural policy conditions.

The slowness of the changes, the continuity of the disadvantages and the growing economic gap between rich and poor countries has given rise to a number of reform ideas, which go far beyond the usual criticism and should be given serious examination.

The Implications of Globalization Process for Africa

Economic growth is not the only dimension of globalization but there are also political, ideological and cultural ones. As a result of deregulation and privatization Indian economy has recently showed faster growth (6.5 percent of GDP). China's entry into the global system and at the same time growth of its

GDP by 8 percent shall not cloud the effects of globalization on other dimensions of human life such as political, ideological etc. The debate about globalization has increased in intensity in the past few years, because of two shock events in the world economy: the Asia crisis of 1995 and the world-wide financial markets crisis of 1998 (Van Der Loop; 2000).

The global economy has passed through two main developments in the past two decades, i.e., a strengthening of cross-border economic relationships, and an ever-growing integration of the international economy. Their economic linkages were also furthered by the greater mobility of goods and services, and of factors of production. Key players in today's global economy are the multinational corporations (MNC's which are about 37,000 in early 1990's), the growing financial and capital markets, and the information and computer technology. In the 1980s and 1990s the following five major international changes have shaped the world economy, i.e., the end of the cold war; the intensification of competition among capitalist powers; the globalization of production and trade; new patterns of development in finance; and new ideological concern such as neo-liberalization, neo-conservation and neo-orthodoxy with elements of macroeconomic stability; a reduced government role in the economy due to deregulation and privatization; and greater openness to the outside world as a result of reduced barriers to trade and a more hospitable approach to foreign capital. This has led to a renewed financial power of IMF and World Bank.

Globalization, which reflects the progressive integration of the world's economies, requires national governments to reach out to international partners as the best way to manage change affecting trade, financial flows, and the global environment. The dominant view of globalization is that closer economic integration will lead to more efficient resource utilization and hence faster economic growth. However, in developing countries it led to uncertainty concerning the possibility of exclusion and that of marginalization. Globalization led to increasing inequalities between as well as within countries.

Globalization has mainly affected trade in manufacturing. Therefore, primary goods producing countries particularly the sub-Saharan Africa and Latin America missed the effects of globalization. Low-income developing countries could not benefit from globalization because of weak infrastructure, low technological capacity and a higher degree of investment risk and uncertainty. All these have reduced the developing countries competitiveness in the world trade/markets.

Globalization has mixed opinions. On the one hand there are authors who stress the positive consequences of globalization, i.e., transfer of technology, raising of productivity and improvement of living standards of some people. On the other hand, there are others who emphasize on the adverse effects of globalization, i.e., marginalization tendencies; increasing in equalities of poor and rich both within and between countries; and generates violence. Moreover as Kumssa (1998) noted, globalization has played two contradictory roles, that is, increased liberalization and competition on the one hand and increased narrow nationalist sentiments / regionalization such as protection of national markets, local identities and ethnicities.

In the globalization process the position of Africa has been characterized by "The Dynamics of Exclusion". Sub-Saharan Africa was virtually economically bankrupt due to the deterioration of the terms of trade (equally in primary commodity production) in 1970s. The oil shock of 1979 led to the imposition of liberalization policies through the Structural Adjustment Police (SAP). The share of SSA in the world trade of manufactured goods has declined from 1.2, to 0.5 and 0.4 percent in 1970, 1980 and 1989 respectively. Similarly the share in primary goods declined from 7.2, to 5.5 and 3.7 percent respectively in the same period. The marginalization of Africa as summarized from Castells (1996) has been a result of the following factors: primary commodities are useless and low priced; markets are too narrow; investments are too risky; labour is not skilled enough; communication and telecommunication infrastructure is inadequate; politics is too unpredictable; and government bureaucracies are inefficiently corrupt.

Africa is a marginalized content in the world export for primary commodities. Africa's share compared to that of Asia in export market for primary commodities declined in 1990s than it was in 1970s. It is also possible to see that Asia replacing

Africa. For example, Cocoa and Coffee declined by 20 and 10 percent for Africa and increased by 20 and 6 percent for Asia respectively in the period between 1970 and 1993.

Africa has become less competitive in the world economy today than in the past. Africa's exports are less diversified today as compared to the past. Africa's exports are less diversified, in 1994 than in 1990. Algeria, who exported 76 different products in 1994 as compared to 49 products in 1980, is the only exception within countries.

Thus Africa is advised to counteract this new international context development by multilateral diplomacy according to Castells (1996) through the OAU and UNIDO; regional economic integration through organizations such as southern Africa Development Community (SADC), the common market for Eastern and Southern Africa (COMESA), and the Economic Community of West African States (ECOWAS); and undertaking political and economic liberalization at both state and popular levels.

CONCLUSIONS

The self-interests of the industrialized nations in trade policy still hinder further reform steps. Whereas many developing countries in recent years have opened their markets, including in the agricultural sector, under the pressure of the WTO negotiations and structural adjustment programmes, the North still keeps its markets closed and distorts the market even more by huge subsidies to promote its own agricultural sectors.

Double standards in the theory and practiced of free trade is a common feature of all round negotiations. The industrialized countries have often failed to practice what they preach. Most often they abandon the comparative advantage argument and protected their own domestic industries and exercised monopolies power over intellectual property. The TNCs are major source of technologies and they control these techniques through patents, investments, cross-licensing and services.

Arguments related to globalization, free trade, liberalization, tariff reductions, tariffication of non-tariff barriers, special safeguard, market access, subsidies reduction to domestic producers, export subsidies reductions and pressures from WTO, World Bank and IMF are all in the final analysis kept developing countries trade open.

The loopholes in trade rules and tricks used to block the agricultural exports of developing countries have slowed the rate of changes toward fair trade, allowed the continuity of the disadvantages and increased the economic gap between rich and poor countries. All these ideas and practices develop a new curiosity of the disadvantages and increased the economic gap between rich and poor countries. All these ideas and practice develop a new curiosity and an urgent fresh examination of trade policies and organizations. Developed countries than simply making rhetoric about development of developing countries, it is the time for them to take some real steps to words it. Developing countries also need the collective self-reliance and unity of thought and action as the option to the economies that are being marginalized in an ever-changing world.

7

Problems of Agricultural Labour

INTRODUCTION

Over the last fifteen years the condition of the rural masses has deteriorated considerably. The survival of agricultural labour, who are a third of the rural population, depend primarily on employment. The Economic Survey of 2004 and 2005 has noted how rural employment has declined from 60 per cent to 57 per cent of all those employed in just one year. We have been noting all along how the days of work of agricultural labour have come down from 122 in the eighties to about 72 now. But what is significant is that a recent study of Punjab and Haryana's Green Revolution areas shows that the decline is sharper where mechanised and corporatised agriculture, in keeping with the views of our present planning commission, is being practiced.

In other words, working days are being reduced on account of the policies enunciated and pursued by the Indian ruling classes for their immediate advantage under the cover of WTO prescriptions. The Indian ruling classes are working to tirelessly dispossess petty producers like marginal farmers, weavers, village craftsmen and agricultural labour. To do that they are first attacking their jobs, and then they will take over their meagre assets using the multinationals with their eye on the huge Indian middle and upper class market as a cover for their predatory policies.

This is evident in agriculture already. It is through the introduction of labour saving machinery, the use of pesticides and shifting from grain production to cash crops, that landlords are squeezing out higher profits by reducing jobs and increasing the work load. The point to note is that while they are themselves under pressure because of the cutbacks in subsidies and through the competition of foreign produce, they are not prepared to side with the mass of peasants and agricultural labour. On the contrary, they are squeezing the miserable income of agricultural labour and attacking even their subsistence.

Moreover between 1991 and 2001 some 3 crore 30 lakh peasants lost their land and have entered the ranks of landless and migrant labour. Every year some 30 lakh such people join the ranks of agricultural labour reducing the possibility of work even further. Also the practice of using women and child labour at cheaper rates further complicates the issue.

Under these conditions we are left with no alternative but to ask the state to ensure the right to life enshrined in the Indian Constitution by guaranteeing rural employment. This is necessary as the rate of growth in industry and private enterprises cannot cope with the massive unemployment resulting from the ruin of the working peasantry, craftsmen and small-scale producers by WTO dictates that the government by its policies has generated. It is its duty to ensure an alternative source of employment.

Another feature that accompanies the shift of government favour from price controls and subsidies to free trade and profiteering is holding the consumer to ransom and squeezing every penny out of him. Even those below the poverty line have not been spared. The prices of rice and wheat, both for APL and BPL categories, have been doubled by the previous NDA government without an increase in earnings. This incapacity to buy has resulted in a glut in government-held food stocks which were sold to the USA as cattle feed at Rs. 4.30 per kg while Indians were expected to pay between Rs. 5.80 to Rs. 7.90 for the same grain.

The present government too is trying to avoid its responsibility towards the PDS. Every day we see news items about government held grain being sold on the black market as an excuse for dismantling the PDS. The union budget also has targeted the Food Corporation of India that is the single buyer of grain for the government. There is a move to multiply buying agencies now, which is likely to discredit the PDS even further and lead to its collapse. This is not acceptable to us. We demand the government to ensure a properly functioning PDS on the Kerala pattern and the provision of rationed goods at BPL prices to all agricultural labour families. Let the government understand that with the rise in price of all items, particularly in the last three months of NDA rule, many agricultural labour families are able to afford only one meal a day. So adequate food-for-work programmes and BPL cards to all agricultural labour families have become necessary for their survival.

Then there is the crucial question of providing the basic infrastructure for agrarian production. This includes subsidies for fertiliser, minimum support prices and cheap or even free electricity. The dismantling of the state electricity companies will lead to people being forced to do without it in the villages because it will become unaffordable for the majority of those living there. The effect will be similar to that of the increase in price of foodgrains on the PDS. Only it will affect rural production more, as a considerable amount of water for irrigation is provided by electrically operated tube-wells. Craftsmen and petty producers in the villages too, like power loom weavers, many of whom come from agricultural labour families, will lose their sources of income with electricity costing three to four times the present price. So we demand that the rural poor with holdings below one hectare and agricultural labour, be given free electricity to give them 'a level playing field' to compete in. In fact in states like Punjab, which provide free electricity to landlords, the SEB continued to send agricultural labour exorbitant bills and forced us to fight for free electricity for agricultural labourers as well, which they only agreed to implement partially.

The question of electricity is also linked to the question of water. The poorest sections, often debarred from community wells on the basis of caste, rely almost completely on government water supply. But, as in UP, state governments are dismantling the Jal Nigams and handing over their tube wells to village panchayats which have neither the funds nor the political will to operate them. We demand that the state governments take the responsibility to provide proper drinking water to the rural poor. This retreat from providing a proper water supply to the poorest sections will affect both farmers and women of agricultural labourer families who work side by side with the men. They are employed more often than the men for lower wages. So if they have to spend more time to fetch water, their capacity to earn and support their families will be severely restricted.

Finally, even when it comes to rural credit, it is unavailable to those who need credit most, agricultural labour. They are forced to borrow money as subsistence loans, for marriages and deaths. Both social and economic necessities force them into debt and often into bondage. The government should address this question with utmost seriousness so that agricultural labour are able to go through the different stages of life without being reduced to dependence as they are traditionally. The best way to do this is for the government to institute subsistence, marriage and death loans without interest to allow agricultural labour to live as human beings and not as attached labour living in near-animal conditions. But if one looks at the present budget, one sees that the government has totally ignored this section in its perspective of rural credit. Worse, it has chosen to provide more credit to those sections that take bank credit to run money-lending businesses or even to attach migrant labour to themselves as bonded labourers. This lacuna in the government's perspective must go. And, it can go only when the mass of agricultural labour, together with farmers and the working class, unleashes struggles for their rights.

This finally brings us to the question of assets. One of the most pernicious developments of the opening up of agriculture to the corporate world is the dispossession of the rural small producer and the rural poor. Not only are their small garden plots up for grabs, so are their house-sites. The women are even more insecure than animals. If they as much as demand the minimum human treatment they can be forbidden to use the village fields to defecate and urinate in. They are subject to rape and murderous attacks. And today, the extra-judicial powers of the rural rich have reached such proportions that inter-caste marriages are often punished by lynching young couples. Without giving these people land, house-sites and lavatories at least, we cannot expect them to live as citizens and as human beings.

The new forest ordinance passed by the BJP-led NDA government in 2002 has dispossessed nearly 16 lakh tribal people as encroachers on land they have lived in for generations. This has increased the oppression of forest officials and made the tribal people insecure. Worse, it has increased the number of migrant labour and made the job situation worse for the rural masses. The forest policy that was drawn up in the interest of multinationals and corporates was successfully opposed by the mass of tribal people in Kerala and by the governments of West Bengal and Tripura. While there is a promise from the government to stop evictions, forest officials and vested interests continue to harass forest dwellers. Much more vigilance will be needed to protect them.

In these conditions, it is not an exaggeration to say that under the neo-liberal market life has become unbearable for the rural poor and especially agricultural labour who are the most oppressed and exploited among them. They cannot survive without struggle against their economic exploitation and against social oppression. And their struggles will not succeed without a perspective and an organisation to ensure it is implemented. As such, they are that section of our society that most needs an agrarian revolution. So it becomes the primary duty of our Party to organise them as the crucial link in the worker-peasant alliance and to unleash its momentum with their full participation.

AGRICULTURAL LABOURERS : CONDITION OF LIVING

Corresponding to the controversy on the nature of transition in the Indian economy a debate has been raised, of late, in respect of the historical genesis of the growth of a class of agricultural wage workers in India. To account for this phenomenon a number of theories have been advanced. Out of these a few of them have been associated with controversy in relation to the origin and growth of these categories of labour force. At one level the origin of agricultural wage labour has been traced back to the pre-capitalist labour-employer relation. However the dynamics of colonial capital penetration in introducing new property relation have been ignored.

At the one end, the reason for the growth process of agricultural labour category is ascribed to the pernicious aspect of the British colonialism bound by imperialism in disintegrating the traditional Indian agrarian society. Again in recent years attempts have been initiated by the academics in equity the rapid growth of agrarian wage workers with that of development of capitalist relation in agriculture. The spatial variation of the category is equated with the uneven development in the growth process of capitalism. In essence they provide evidence of capitalism in agriculture by measuring the wage labour employment.

To examine some of the above mentioned issues the structure of the economy in Orissa through a case study of two villages has been taken: one a wet village and the other a dry village located in Sambalpur district. The study begins with the agricultural labour situation during the last three censuses in Orissa. It is discerned that there has been a growth of agricultural wage worker category over time at the aggregate and percentage level as well. Spatially both advanced and backward districts are having higher proportion of agricultural labourers to the state average. Hence contrary to the view point expressed in many quarters no correspondence between the level of development and growth of agricultural wage workers has been found.

The agrarian structure during British colonial periods indicates that the process of change centred around revenue settlements, commercialisation of agriculture, in addition to the communication net work. They contributed to the restructuring of the economy, though the restructuring was not to the extent of disintegration of the feudal hold over the area. Capital penetration failed to transform the socio-economic structure in its own image. Rather it compromised with the existing pre-capitalist production processes. Hence internal colonisation, caste dominance, high land revenue, Bethi and Begar system, extraction of surplus through extra economic coercion were identified as some of the devices through which the tenants, share croppers, and landless agricultural labourers were exploited. There was little change in the old pattern of social relation of production.

Post colonial Orissa witnessed a series of state interventions aimed at developing the economy on capitalist lines, princely state were integrated with the Indian Union and land reform measures were initiated to give a blow to the feudal land relations. Later on, technological intervention i.e. HYV seed fertiliser and irrigation water were introduced. The post colonial state intervention, capital penetration and its consequences had been reflected in the structure of the villages that were studied. The significant structural changes as observed in the wet village displayed the growing importance of the non-cultivating peasants and the stability of the middle peasants households over time. The rich peasants as a special group is weak one in transforming the forces of production and production relation.

The structure that emerges from the dry village depicts the strengthening on the non-cultivating peasants households and the disintegration of the middle peasants. However, the capitalist - rich peasant structure does not appear to have gained in economic strength. Wherever capital has gone it has restructured the feudal economy. In the process of the operation of capitalistic law of motion there is a differentiation among the peasantry in the agrarian economy. However, capitalism in an excolonial country like India has developed with a compromise with pre-capitalistic mode of production. Hence there is a simultaneous co-existence of advanced productive force along with backward production relation.

The differentiation in the peasantry has been due to the state intervention in the productive forces and legal structures. More particularly, changes in the legal structure in respect of common property resources introduced during the British period and continuing till date in it minor amendments have further differentiated the peasantry each of these developmental processes have two dimensional effects: At one level it integrates and at the other it disintegrates. In India the latter process is at work and not the former one. Because the 'developmental projects' have dislodged a substantial section of population without re-integrating them in the economy. More particularly poor peasants and agricultural labourers have been marginalised over time.

This marginalised section is not purely on account of the capitalist differentiation of the peasantry. The dynamics of structural change are not strong enough to dissolve the feudal remnants nor it can bring about a differentiation of the peasantry in its pure form by introducing capitalist rich peasant culture.

Summing up all these findings, it is found that the presence of middle peasant economy provides a narrow base for sustained expansion in the economic activity in the wet village. This has been manifested in the rigidity in the cropping pattern, low level of farm mechanisation, narrowness of the home market etc. Hence there is no clear evidence of the acceleration in the economic growth leading to a major structural change. Hence the difference between the wet and dry village is rather marginal. Similarly the production relation that is existing in these two villages are not qualitatively different.

The employer-employee relations, the mode of wage payment, wage rate, quantum of employment, household budget, incidence of indebtedness etc. are backward and archaic as in the dry village.

In the ultimate analysis, the above mentioned historical events rather than the growth of capitalist relations in agricultural production alone have contributed for the immense growth of agricultural labourers. Therefore, the nature, growth and dynamics of agricultural labour have more to do with the socio-economic context in which growth is taking place. This is more important than the growth itself.

8

Child Labour in Agriculture

INTRODUCTION

Of nearly 250 million children engaged in child labour around the world, the vast majority-70 percent, or some 170 million-are working in agriculture. Child agricultural workers frequently work for long hours in scorching heat, haul heavy loads of produce, are exposed to toxic pesticides, and suffer high rates of injury from sharp knives and other dangerous tools. Their work is gruelling and harsh, and violates their rights to health, education, and protection from work that is hazardous or exploitative.

According to the International Labour Organization's new report on child labour, the number of children working in agriculture is nearly ten times that of children involved in factory work such as garment manufacturing, carpet-weaving, or soccer-ball stitching. Yet despite their numbers and the difficult nature of their work, children working in agriculture have received little attention compared to child labour in manufacturing for export or children involved in commercial sexual exploitation.

In investigations in Egypt, Ecuador, India, and the United States, Human Rights Watch has found that the children working in agriculture are endangered and exploited on a daily basis. Human Rights Watch found that despite the vast differences among these four countries, many of the risks and abuses faced by child agricultural workers were strikingly similar.

In Egypt, Human Rights Watch examined the cotton industry, Egypt's major cash crop, where over one million children work each year to manually remove pests from cotton plants. In Ecuador, where nearly 600,000 children work in the rural sector, the organization investigated conditions for children working in banana fields and packing plants. In the United States, Human Rights Watch examined conditions for the estimated 300,000 children who work as hired labourers in large-scale commercial agriculture, planting, weeding, and picking apples, cotton, cantaloupe, lettuce, asparagus, watermelons, chillies, and other crops. In India, Human Rights Watch looked at bonded child labourers working in agriculture as part of a larger study of bonded child labour. There are as many as 15 million bonded child labourers in India, most of whom are Dalits (untouchables) or lower caste. More than half, and possibly as many as 87 percent of these bonded child labourers work in agriculture, tending crops, herding cattle, and performing other tasks for their "masters."

The International Covenant on Civil and Political Rights (ICCPR) states, "Every child shall have . . . the right to such measures of protection as are required by his status as a minor, on the part of his family, society and the State." The Convention on the Rights of the Child provides that children-all persons under eighteen "unless under the law applicable to the child, majority is attained earlier"-have a right "to be protected from performing any work that is likely to be hazardous or to interfere with the child's education, or to be harmful to the child's health or physical, mental, spiritual, moral or social development." All states parties to the Convention-every government in the world except for the United States and Somalia-are required to "undertake all appropriate legislative, administrative, and other measures for the implementation of the rights recognized in this Convention."

ABUSES OF CHILD AGRICULTURAL WORKERS

Ages and Hours of Work

Child agricultural workers often begin work at early ages, and may work twelve or more hours a day. In India, bonded child labourers as young as eleven often work sixteen or seventeen hours a day, typically beginning at 5 or 6 a.m., and continuing until 9 or 10 at night. Some are expected to work 365 days a year. In Ecuador, children working in the banana sector typically start work at ages ten or eleven, though some begin as early as age eight. Although some work only five hours a day, Human Rights Watch found that the vast majority worked between nine and thirteen hours a day.

In the United States, children interviewed by Human Rights Watch began working in the fields as early as age twelve. They routinely worked twelve-hour days, and during peak harvest season, sometimes worked fourteen hours or more. Children may begin working as early as 4 a.m., and may spend two hours or more each morning and evening travelling to the fields where they work.

Children working in cotton pest control in Egypt are typically between the ages of seven and twelve. For periods of up to ten weeks each year, they work eleven hours a day, seven days a week.

Pesticide Exposure

One of the greatest threats to the health of child agricultural workers is exposure to pesticides. In Ecuador, Egypt and the United States, children reported working in freshly sprayed fields, and even working in fields while they were being sprayed. Children interviewed reported symptoms of exposure including headaches, fever, dizziness, nausea, rashes and diarrhea. In severe cases, pesticide exposure can lead to convulsions, coma and death. Long term effects also include cancer, brain damage, sterility or decreased fertility, and birth defects.

Child agricultural workers are often not told of the dangers of pesticides, or how they can protect themselves. In the United States, not one of the children interviewed by Human Rights Watch had received training about the dangers of pesticides, safety measures, or what to do in case of exposure. Some did not even know what pesticides were. "Pesticides? Was that the medicine they put on [the crops]?" said one girl. "No, I don't know anything about that."

In Ecuador, 90 % of child banana workers interviewed by HRW stated that they continued working while fungicides were sprayed from airplanes flying overhead. They described trying to protect themselves by hiding under banana leaves, covering their faces with their shirts, or placing banana cartons on their heads. One boy said, "I went under the packing plant roof until the [fumigation] plane left-less than an hour. I became intoxicated. My eyes were red. I was nauseous. I was dizzy. I had a headache. I vomited."

In all three countries, children may participate in applying pesticides to crops. In Egypt, children sometimes operate motorized pumps that saturate cotton plants with pesticides. Half a dozen children may help carry the pump's long hose, becoming heavily contaminated with pesticides in the process. In Ecuador, children frequently handle pesticide-treated plastics used to protect banana stalks in the fields, and spray fungicides onto bananas being prepared for shipment in packing plants. Many of the children reported to Human Rights Watch that they did not use any protective equipment, even gloves, when handling the chemicals.

In Egypt, two of the five pesticides recommended for use by the government are categorized as "highly hazardous" by the World Health Organization. In Ecuador, one pesticide in common use is the subject of over one hundred lawsuits from around the world, alleging that the chemical is responsible for serious birth defects, including cleft palate and being born with no eyes.

Pesticide risks are particularly acute for children. Because their organs are still developing, they are less able than adults to expel toxins from their body. Their breathing rate is much higher than adults, and they have more skin surface per unit of body weight than adults, allowing them to both breathe in and absorb higher concentrations of toxic chemicals.

Injuries and Disabilities

Children working in agriculture suffer high rates of injuries. They frequently suffer cuts from sharp knives and falls from ladders. They risk back injuries from hauling heavy loads of produce. They may be crushed or maimed by tractors and other heavy equipment.

In the United States, agriculture is second only to mining for occupational fatalities. Child farmworkers make up only 8% of children who work in the United States, yet account for 40% of work-related fatalities among minors. An estimated 100,000 children suffer agriculture-related injuries each year in the United States.

In Ecuador, children commonly use sharp knives and machetes to cut yellow leaves off banana plants, and curvos-short, thick, crescent-shaped blades with wooden handles-to cut bananas off their stalks, to cut plastic bags used to cover banana stalks, and for other tasks. A quarter of the children interviewed by Human Rights Watch had cut themselves with these sharp tools at least once.

In Ecuador, boys also haul heavy loads of bananas from the fields to packing plants using a harness system attached to an iron pulley riding on cables. One boy, now fourteen, reported that when he was ten years old, he would pull twenty banana stalks (each weighing 50 to 100 pounds) at one time from the field to the plant. The distance from the field to the plant was two kilometers; he would make five or six trips per day, each trip taking about one hour. Dragging the heavy fruit can cause back injuries to children. In addition, stalks of bananas or wheels on the cables sometimes fall off and strike children, causing serious injuries.

Water and Sanitation

Child agricultural workers often must work in the full sun in temperatures exceeding 100 degrees Fahrenheit or 40 degrees Celsius. Under such conditions, health experts recommend that workers drink two to three gallons of water a day. Without adequate drinking water, workers run the risk of devastating dehydration and heat illnesses that can cause death or brain damage. However, many child agricultural workers do not have access to water, or are provided with water contaminated with bacteria or pesticides.

In Egypt, some children worked from 7 a.m. until 6 p.m, but reported that only during two breaks during the day did their foremen allow them to drink. In both the United States and Ecuador, some children reported that a lack of potable water forced them to drink from canals that drain excess water from the fields. These canals are often contaminated with pesticides, fertilizers, bacteria and human waste. Children may also be forced to use their earnings to purchase water, soda or beer from their employers at inflated prices.

One boy in the United States reported, "We had to share water from one big jug. It wasn't enough. You couldn't drink as much as you wanted.... An old man took us there [to the field] in the morning, set us up, then would come back in the afternoon to pick us up. If you ran out of water, if you passed out, tough."

In the United States, agricultural employers are required by law to provide toilet facilities, drinking water, and water for handwashing. However, nearly all of the children interviewed by Human Rights Watch reported working in fields or orchards where one or more of these requirements were not met.

Lack of toilet facilities contributes to the spread of parasitic infection among workers. It is particularly dangerous and humiliating for girls, who may be forced to choose between public urination-more obvious and awkward for females-and urinary retention, which can cause severe discomfort and urinary tract infections. Some workers also report limiting their fluid intake to prevent urination, at severe risk to their health.

Lack of water for handwashing also is frequently unavailable. In addition to being unsanitary, unwashed hands virtually guarantee that pesticides will be ingested when workers eat their lunch.

Ill-Treatment and Sexual Harassment

In Egypt, Human Rights Watch found that children working in cotton fields are routinely beaten. Children typically gauged the leniency of a foreman by the severity and frequency of the beatings he administered. One ten-year-old boy worked under two foremen. "One of them I hate; the other one I like. The one I hate used to beat and kick me whenever I missed a leaf. The other one beats and kicks me lightly."

Severe maltreatment caused some children to quit work entirely or seek employment under the supervision of a different foreman. A nine-year-old Egyptian girl described a steady process of attrition from her work group. "The last work group I was in started with twenty-two [children], but you know, children don't like to be hit, so they turn up in another foreman's group. Our group ended up with twelve."

Bonded child agricultural workers in India also frequently experience physical abuse. An eleven-year old told Human Rights Watch, "I do not like work; it is hard and there is no time limit. When I'm sick, the master won't let me stay home. If I try to take time off he will scold me and beat me and take me back to the fields. Sometimes he beats me because he says I am working slowly." A girl who began working in bondage at age nine reported that her master beats her, and yells and curses at her.

In both the United States and Ecuador, girls working in the fields reported sexual harassment by their supervisors. In Ecuador, a twelve-year-old reported that when she bent down to pick up plastic bags, her boss would say, "There is a good place to stick my balls." An adult packing plant worker in Ecuador told Human Rights Watch that both supervisors and other male workers grabbed the breasts and buttocks of girls.

In the United States, girls are routinely subjected to sexual advances by farm labour contractors and field supervisors. If they refuse, they-and members of their family-face retaliation in the form of discharge, blacklisting and even physical assault and rape. An eighteen-year old said, "Everyone is scared to say anything because they threaten them. If they say something they will lose their job."

Impact on Education

In Egypt, cotton pest control work usually takes place during the summer months when children are not attending school. However, in other cases, agricultural work can have severe consequences for children's education. Long hours of work cause children to miss classes and leave them too tired to study. Eventually they fall behind and frequently drop out completely. In Ecuador, the majority of children with whom Human Rights Watch spoke had quit school before the age of fifteen. Of those still in school, several explained that they often missed school to work. In the United States, only 55 percent of farmworker children in the United States finish high school. Of the dozens interviewed by Human Rights Watch, nearly every one had dropped out of school for at least one extended period of time.

In India, many bonded child labourers working in agriculture have never been to school. Once they are bonded-often at age eight or nine-their long hours of work frequently make schooling impossible. One thirteen-year old boy who worked from 5 a.m. to 6 p.m. tending sheep and picking mulberry leaves was enrolled in a non-formal school run at night, but told Human Rights Watch, "When I have a lot of work, I am not allowed to come to the school. . . [t]he owners complain I am not earning enough."

Wages

Child agricultural workers work for very low wages, and are often paid less than their adult counterparts. In Egypt, children working harvesting cotton in one region earned US $1.08 per day, while men earned US $1.63 per day and women earned US $1.36. Children working in cotton pest control-a type of work typically not done by adults -earn between US $0.68 and US $0.95 per day.

In Ecuador, the legal minimum wage for a banana worker is US $5.85 per day. Adult workers interviewed by Human Rights Watch earned, on average, approximately US $5.44 per day, while children averaged only US $3.50, 60 percent of the legal minimum wage for banana workers.

In the United States, the legal minimum wage is $5.15 per hour. Approximately one-third of the child farmworkers interviewed by Human Rights Watch earned significantly below minimum wage, and some were paid as little as $2.00 per hour. Some children were instructed by their supervisors what to say if they were approached by a government inspector: "If they ask how much I pay you, say $5.50" or "say $6.00."

In India, children bonded into agricultural labour are working to pay off a debt. Parents or other relatives promise the labour of the child to an employer in exchange for a sum of money. The children then spend long hours over many years in an attempt to pay off these debts. Due to high interest rates charged and abysmally low wages, they are usually unsuccessful. Often the yearly or monthly sum deducted from the loan is not even one-quarter or one-third of the prevalent daily wages, let alone the legal minimum wage.

One thirteen-year old Indian boy was bonded for two years in exchange for a 7,000 rupee (US $143) advance given his family in order to repay another loan. The boy told Human Rights Watch that for each day of labour, a "wage" of 20 rupees (US $0.40) is applied against the loan. At a rate of 20 rupees a day, the loan would be repaid in less than one year, but inflated interest rates will ensure that the boy labours for at least twice that period of time.

Laws and Their Enforcement

In some countries, such as Egypt and Ecuador, national laws set appropriate limits on both the ages and hours at which children can work and, if enforced, would help to protect children from abusive labour conditions. In other cases, like the United States, laws are clearly inadequate and fail to protect children from violations of their rights.

In both Ecuador and Egypt, national law prohibits children from working before the age of fourteen, but allows twelve and thirteen-year olds to work as apprentices as long as the work does not threaten their health or education. Egyptian children and Ecuadorian children under the age of fifteen are not allowed to work for more than six hours a day. In Ecuador, children between fifteen and eighteen are allowed to work for seven hours a day. Other regulations and legal provisions are also intended to protect working children, and Ecuadorian law explicitly prohibits hazardous work, including handling toxic substances and tasks that are "considered dangerous or unhealthy."

Unfortunately, both governments fail to enforce these laws, placing hundreds of thousands of children at risk and violating their rights to health and development.

In the United States, child agricultural workers receive fewer legal protections than other working children. U.S. labour law allows children in agriculture to work at younger ages, for longer hours and under more hazardous conditions than children in other jobs. While the law allows children as young as twelve to work unlimited hours in agriculture, children in other occupations cannot work before age fourteen, and can only work three hours on a school day until age sixteen. In addition, even limited protections in existing law are not adequately enforced. Only a tiny fraction of child labour violations are ever uncovered by the Department of Labour, and penalties are typically too weak to discourage employers from using illegal child labour.

Bonded labour, whether in agriculture or any other area, is illegal under India law and by law, bonded labourers are to be released and rehabilitated and their employers prosecuted. The law also prohibits children under age fourteen from handling pesticides or insecticides. However, Human Rights Watch has found that the Indian has failed to adequately enforce its own laws; even where children are identified and released from bondage, often through NGO's intervention, they often receive no rehabilitation and their employers not penalized.

Egypt, Ecuador and the United States all are party to the ILO Convention 182 on the Worst Forms of Child Labour. The Convention prohibits "work which, by its nature of the circumstances in which it is carried out, is likely to harm the health, safety or morals of children." Each country is expected to determine what constitutes such work, although the ILO's Recommendation 190 recommends that such work include : (a) work which exposes children to physical, emotional or sexual abuse; (b) work underground, under water, at dangerous heights or in confined spaces; (c) work with dangerous machinery, equipment and tools, or which involves the manual handling or transport of heavy loads; (d) work in an unhealthy environment which may, for example, expose children to hazardous substances, agents or processes, or to temperatures, noise levels or vibrations damaging to their health; or (e) work under particularly difficult conditions such as work for long hours or during the night or work which does not allow for the possibility of returning home each day."

Based on Human Rights Watch's findings, child agricultural workers face many of the conditions outlined by the ILO as work likely to harm their health and safety, in particular, work with dangerous machinery, equipment and tools; work in an unhealthy environment, including exposure to hazardous substances, notably pesticides; and work for long hours. In addition, girls may also face the danger of sexual abuse.

The conditions for child agricultural workers also violate the Convention on the Rights of the Child, which has been ratified by every government except the United States and Somalia. The Convention states that children have the right "to be protected from economic exploitation and from performing any work that is likely to be hazardous or to interfere with the child's education, or to be harmful to the child's health or physical, mental, spiritual, moral or social development." (Article 32) Article 24 recognizes the right of all children to a high standard of health, article 28 recognizes the right of all children to education, and article 34 states that children should be protected from all forms of sexual exploitation and sexual abuse. Finally, article 3 of the Convention states that "In all actions concerning children . . . the best interests of the child shall be a primary consideration."

Recommendations

The extraordinary numbers of children working in agriculture -some 170 million worldwide- and the severe abuses they endure demand that governments, employers and trade unions prioritize protections for child agricultural workers as part of their strategies to end child labour.

Human Rights Watch recommends that all governments:

(i) Ratify the ILO 182 Convention on the Worst Forms of Child Labour, if they have not already done so, and ensure that their national laws are consistent with the Convention and Recommendation 190. National laws should explicitly prohibit all individuals under the age of eighteen from using dangerous tools, from hauling heavy loads, from working long hours, from handling pesticides, and from being exposed to pesticides in the workplace. National laws should also explicitly prohibit sexual harassment in the workplace, establishing separate and more stringent penalties for cases in which the victim is a minor.

(ii) Take measures to ensure the effective implementation of Convention 182. Such measures should include the allocation of resources to provide for a sufficient number of labour inspectors targeting child labour in agriculture, proactive monitoring, and unannounced on-site inspections.

(iii) Ensure that restricted-entry intervals (REIs)-the time after pesticide application when entry into the treated areas is banned or limited -are clearly established in government regulations and vigorously enforced. Special REIs for children should be established, taking into consideration the greater risks they face from exposure to toxic chemicals.

(iv) Ensure that all workers, including children, receive full information and training from their employers about occupational illnesses and injuries related to agricultural work, including those associated with exposure to pesticides. Such training should be conducted regularly and be understandable by children. Employers should also provide all workers, including children, with appropriate protective equipment and train them in methods of protecting themselves from workplace hazards.

(v) Ensure that children and their families are aware of the rights of children.

(vi) Undertake comprehensive surveys to determine the scope and scale of child labour in the agricultural sector, the number and nature of injuries or illnesses suffered by children working in agriculture, and disaggregate the data by sex and age.

(vii) Ensure that all children receive free and compulsory primary education. School fees and other associated costs of education, including costs for books and uniforms, should be waived, or scholarship programs developed for children whose families are unable to afford them. Special educational or vocational programs should be developed for child farmworkers who have dropped out of school.

(viii) Violators of child labour laws should be sanctioned to the fullest extent of the law. Governments should consider increasing fines for child labour violations and dedicating a portion of the fine to the rehabilitation of child workers.

(ix) Ensure that child agricultural workers who labour in hazardous conditions in violation of ILO Convention 182 and suffer workplace accidents or illnesses are provided with free health care.

(x) Establish mechanisms to monitor the treatment of children working in agriculture and ensure that effective complaint mechanisms are available to children and their families.

Human Rights Watch recommends that all corporations who employ children or with suppliers who employ children:

(i) Ensure that children's human rights are respected on all directly owned and supplier farms and plantations by adopting effective monitoring systems to verify that labour conditions on these facilities comply with internationally recognized child labour standards and relevant national child labour laws, reporting annually on compliance, and, where the facilities fall short, by providing the economic and technical assistance necessary to bring them into compliance.

(ii) Ensure that pesticides potentially harmful to children are neither sanctioned for use nor applied in practice on directly owned or supplier farms and plantations.

(iii) Immediately turn over to the appropriate national authorities any information regarding violations of national child labour laws or international child labour standards on supplier farms and plantations.

(iv) Provide adequate support for underage workers, as defined by the ILO Minimum Age Convention, to attend school or an appropriate academic alternative in lieu of working.

9

Rights of Women Agricultural Workers

INTRODUCTION

In many regions of the world, women make up a considerable portion of the agricultural labour force, as men often migrate from rural areas and/or are employed in non-agricultural occupations (a phenomenon referred to as "feminization of agriculture"). This chapter examines the rights of women agricultural workers, in relation to both access to employment and treatment during employment.

Agricultural labour rights are mainly determined by labour law, and particularly by two broad groups of norms: those concerning all workers, both male and female (minimum wage; safety and hygiene; trade union rights; etc.), and those specifically concerning women (non-discrimination; maternity leave; "protective" legislation; etc.). The focus here is on the latter. While some labour-law issues are relatively uncontroversial (e.g. non-discrimination), others are debated. For instance, "protective" legislation prohibiting women from working in certain occupations or at night, enacted to protect women workers, limits women's freedom to choose their occupation and may hinder their access to employment. On the other hand, where the bargaining power between employer and employee is particularly unbalanced, allowing women to choose may leave them unprotected (e.g. on night work). In reviewing labour legislation, as it applies to agricultural workers, it must be remembered that in many countries (especially developing countries) these rules are not applied to a large sector of the economy, the informal sector.

Beyond labour law, other norms are also relevant. For instance, in some countries family law allows the husband to interfere in his wife's occupation, e.g. by requiring his consent for her signing employment contracts and by allowing him to terminate her contract if he deems it necessary for the fulfilment of her family obligations.

The law on women's labour rights rarely refers directly to agricultural workers. It more commonly relates to urban occupations (secretaries, civil servants, etc.), especially in developing countries, where access to courts for rural women is usually very limited. However, the principles affirmed in the cases quoted in this chapter (e.g. non-discrimination in the workplace) apply also to agricultural workers.

Discrimination may be difficult to detect where it is indirect or where women's employment opportunity and treatment are affected by entrenched socio-cultural attitudes and unequal access to education and training rather than by formal legislation. A hidden form of discrimination is maternity benefit payment by the employer, rather than by social security institutions; this raises the cost of women's labour (due both to the time lost during maternity leave and to maternity benefit payment), fostering discrimination in women's access to employment, particularly where fertility rates are high (as in most developing countries). As for socio-cultural practices, in many areas women's participation in formal employment is hindered by their primary responsibility for domestic work and child care. Where women are employed, their workload is very heavy, as they perform their domestic responsibilities in addition to formal employment.

In examining the labour rights of women agricultural workers in each of the covered countries, the following outline will generally be followed: applicability of labour legislation to the agricultural sector; norms concerning access to employment (both under labour law and under family law); norms concerning treatment (remuneration and other terms and conditions of employment); maternity protection; norms on social security; and sanctions. Where available, information on the actual implementation of the norms referred to is also included.

RELEVANT INTERNATIONAL LAW

The right to work without discrimination is recognized in the UDHR (arts. 2 and 23), in the ICESCR (arts. 2(2) and 6-8) and in the CEDAW (art. 11). It includes the right to freely choose an occupation, to enjoy a just and favourable remuneration, to work in safe and healthy conditions, and to form and join trade unions. Women have a right to employment opportunities and treatment equal to men, including equal remuneration for work of equal value (UDHR, art. 23(2), ICESCR, art. 7(a) (i) and CEDAW, art. 11). Women also have the right to enjoy special protection during pregnancy and paid maternity leave, and the right not to be dismissed on grounds of pregnancy or maternity leave (CEDAW, art. 11(2)).

As for women's labour rights under the CEDAW, it is worth recalling that the principle of non-discrimination enshrined in this treaty is not limited to state action, and that article 2(e) explicitly envisages the elimination of discrimination against women "by any person, organization or enterprise".

Every woman, as well as every man, has a right to social security in cases of retirement, unemployment, sickness, invalidity and old age (UDHR, art. 22, ICESCR, art. 9 and CEDAW, arts. 11(1)(e) and 14(2)(c)).

Detailed provisions on women's labour rights are contained in several ILO conventions. The Discrimination (Employment and Occupation) Convention 111 of 1958 prohibits sex discrimination in both opportunity and treatment, and provides for affirmative action. The Equal Remuneration Convention 100 of 1951 states the principle of equal remuneration for men and women for equal work or work of equal value. Reference to "work of equal value", besides "equal work", is important for the practical application of the principle. Indeed, since in many countries women rarely hold the same position as men due to cultural stereotypes and unequal access to education, reference to the economic value of the work allows comparisons across occupational categories and industries. The Termination of Employment Convention 158 of 1982 prohibits dismissal on grounds of sex, marital status and absence during maternity leave.

Under the Night Work (Women) (Revised) Convention 89 of 1948 and its 1990 Protocol, women's work at night (defined as a period of at least 11 consecutive hours, including at least seven hours between 10pm and 7am) is prohibited for some industrial occupations (not for agricultural work).

The Maternity Protection (Revised) Convention 103 of 1952 entitles pregnant workers to a maternity leave of at least 12 weeks (with no less than six weeks after childbirth); allows additional leave in case of late delivery or pregnancy-related illness; prohibits dismissal while on maternity leave; entitles women to medical and cash benefits, provided through either compulsory social insurance or public funds; and allows work interruptions for nursing purposes. While this Convention applies to both industrial and non-industrial occupations, states may exempt work in agricultural undertakings other than plantations. The Workers with Family Responsibility Convention 156 of 1981 prohibits discrimination against men and women workers with family responsibilities (e.g. family responsibilities are not a valid reason for termination of employment).

The Plantations Convention 110 of 1958 and its 1982 Protocol protect the labour rights of plantation workers, without discrimination on the basis of sex. The term plantation is defined as "any agricultural undertaking regularly employing hired workers which is situated in the tropical or subtropical regions and which is mainly concerned with the cultivation or production for commercial purposes of coffee, tea, sugarcane, rubber, bananas, cocoa, coconuts, groundnuts, cotton, tobacco, fibres (sisal, jute and hemp), citrus, palm oil, cinchona or pineapple; it does not include family or small-scale holdings producing for local consumption and not regularly employing hired workers" (article 1(1) of the Convention, as amended by the Protocol). State parties may exclude or add categories of agricultural undertakings from the application of the Convention. The Convention contains guarantees as to the recruitment (e.g. the recruitment of the household head does not involve the recruitment of household members), employment contracts, wages (e.g. wages are to be paid directly to the worker), annual paid leave and weekly rest, compensation for injury, trade unions (e.g. workers' freedom of association "without distinction whatsoever"), and maternity protection (maternity leave of at least 12 weeks, at least six of which after childbirth, with additional leave for late delivery or pregnancy-related illness; cash and medical benefits; protection from dismissal during maternity leave; breaks for nursing purposes; prohibition for a pregnant woman to "undertake any type of work harmful to her in the period prior to her maternity leave").

The Migration for Employment (Revised) Convention 97 of 1949 provides guarantees for lawfully migrant workers, without discrimination on the basis of sex.

The principle of non-discrimination is also stated in the 1998 ILO Declaration on the Fundamental Principles and Rights at Work. This declaration reaffirms some fundamental principles and rights to which all ILO member states must adhere by the very fact of their ILO membership, regardless of their ratification of the relevant conventions.

Table 1
Ratification of Some International Labour Conventions

	C 89	*C 97*	*C 100*	*C 103*	*C 110*	*C 111*	*C 156*	*C 158*
Brazil	1957	1965	1957	1965	1965, denounced 1970	1965 No	1995, denounced	1996
BurkinaFaso	No	1961	1969	No	No	1962	No	No
Fiji	No	No	No	No	No	No	No	No
India	1950	No	1958	No	No	1960	No	No
Italy	No	1952	1956	1971	No	1963	No	No
Kenya	1965	1965	2001	No	No	2001	No	No
Mexico	No	No	1952	No	1960	1961	No	No
Philippines	1953	No	1953	No	1968	1960	No	No
South Africa	1950	No	2000	No	No	1997	No	No
Tunisia	1957	No	1968	No	No	1959	No	No

C 89: ILO Night Work (Revised) Convention, No. 89 of 1948 (total 65 ratifications).

C 97: ILO Migration for Employment (Revised) Convention, No. 97 of 1949 (total 42 ratifications).

C 100: ILO Equal Remuneration Convention, No. 100 of 1951 (total 156 ratifications).

C 103: ILO Maternity Protection (Revised) Convention, No. 103 of 1952 (total 40 ratifications).

C 110: ILO Plantations Convention, No. 110 of 1958 (total 12 ratifications).

C 111: ILO Discrimination (Employment and Occupation) Convention, No. 111 of 1958 (total 154 ratifications).

C 156: ILO Convention on Workers with Family Responsibilities, No. 156 of 1981 (total 33 ratifications).

C 158: ILO Termination of Employment Convention, No. 158 of 1982 (total 32 ratifications).

Source: ILO web site (www.ilo.org).

THE AMERICA

REGIONAL OVERVIEW

In the region, the human right to work and to just, equitable, and satisfactory conditions of work, without discrimination on the basis of sex, is stated in the Additional Protocol to the ACHR (arts. 3, 6 and 7). As for Canada, United States and Mexico, the North American Free Trade Agreement (NAFTA) includes a North American Agreement on Labor Cooperation, which states the principles of non-discrimination on the basis of sex and of equal pay for equal work (principles 7 and 8 of Annex 1 to the Agreement on Labor Cooperation).

In most Latin American countries, women may freely enter in employment contracts and dispose of their wage. However, in practice rural women often ask for their husband's authorization before undertaking a job, and quit it if their husband so requires (FAO, 1994). Moreover, some laws explicitly allow the husband to interfere with the employment of his wife, although there is a trend throughout the region to repeal these norms (for an example from Guatemala).

Most countries have constitutional norms and/or ordinary legislation prohibiting sex/gender discrimination on the workplace, either in general or with specific regard to agricultural labour. The equal remuneration principle has been adopted within most national legal systems. However, reference is usually made to equal pay for "equal work" (or similar formulas), instead of the internationally recognized standard of "work of equal value". Overall, a considerable gender pay gap remains throughout the region; for instance, in Paraguay, men earn 31 percent more than women for each hour worked (CEACR (100), 2000).

Maternity leave ranges from 60 days (Bolivia) to 14 weeks (Panama), with a considerable number of countries granting 12 weeks (Belize, Colombia, Haiti, Jamaica, Uruguay). In Venezuela, maternity leave is 18 weeks (Comprehensive Labour Act of 1997). Cash benefits range from 60 percent of the wage (Dominica, Nicaragua) to 100 percent (Chile, Colombia, Venezuela); in the United States, maternity leave is unpaid. In most countries cash benefits are funded by social security institutions (Chile, Paraguay, Venezuela), although in some cases they are paid by the employer (Jamaica) or jointly by social security and the employer (Costa Rica) (United Nations, 2000, updated to 1998). However, there have also been reports of lack of benefit payment in plantations; for instance, in Guatemala, maternity benefits for women plantation workers are paid only in some regions (CEACR (110), 1997).

In plantations, there is a widespread practice of recruiting women as temporary workers, without contract and on piece-work. This non-formalized situation entails the non-application of the protection accorded by labour law, and therefore sex/gender discriminatory practices (FAO, 1994 and 1996).

MEXICO

Article 123 of the Mexican Constitution recognizes the right to work of "every person". The Federal Labour Code of 1970 (applicable to agricultural labour; art. 1 of the Code and art. 123 of the Constitution) prohibits sex discrimination (art. 3).

As for access to employment, both women and men over 16 years (and those between 14 and 16 years if they have the authorization of their parents) can freely enter into labour contracts (Federal Labour Code, art. 23). Women have thus full legal capacity to work. Employers cannot refuse job applications on grounds of sex (Federal Labour Code, art. 133(I)). Provisions banning women from certain types of work were repealed in 1974.

Discriminatory provisions on women's employment opportunities are contained in the civil codes of some states. The Civil Code of Oaxaca states that the wife can hold an occupation only if this does not prejudice her primary responsibility as housewife (arts. 167 and 168). The husband may oppose the employment of his wife, provided that he earns sufficiently for the needs of the family; where the wife resists the opposition of the husband, the dispute is to be decided by courts (arts. 169 and 170).

In other states, similar provisions are expressed in gender neutral terms, with each spouse having the right to oppose the employment of the other (e.g. Civil Code of Aguascalientes; art. 165, the Civil Code of Guanajato as amended in 2000;, art. 168, the Civil Code of Sonora, art. 261).

As for treatment, men and women workers have equal rights and obligations (Labour Code, art. 164). Labour conditions cannot be inferior to the minimum legal requirements, without discrimination on the basis of sex (art. 56). The principle of equal remuneration for equal work (though not for work of equal value) is stated in Article 123(VII) of the Constitution and article 86 of the Labour Code.

However, sex/gender discrimination is in practice widespread. A considerable number of rural woman are employed as temporary workers, without contract and on a piece-work basis, and are thus not protected by labour law (FAO, 1994)[35]. Wage differentials vary considerably from sector to sector; in agriculture and fishing, women's average hourly earnings are 92 percent of men's (CEACR (100), 1998).

Pregnant women cannot be required to undertake heavy and dangerous work (Labour Code, art. 166). With particular regard to agriculture, the Federal Safety, Hygiene and Working Environment Regulations of 1997 contain specific provisions protecting the health of pregnant women. Under article 154, pregnant women cannot be required to work in the operation, transport or storage of teratogenic or mutagenic substances, while under article 155 they cannot use chemical substances (e.g. fertilisers, etc.).

Pregnant workers have a right to maternity leave of at least six weeks before and six weeks after childbirth (with possible extensions); to the payment of the full wage; and to retain their employment (Constitution, art. 123(V) and Labour Code, art. 170). Under the Social Security Act of 1995, maternity benefits equivalent to 100 percent of the worker's salary are paid for 48 days before and 48 days after childbirth by the Mexican Institute of Social Security under a mandatory scheme, provided that the pregnant worker meets specified requirements; if and to the extent to which this norm is applicable, the employer does not have to pay the full salary (arts. 101-103). The Act also provides for a child day-care system (arts. 201-207).

BRAZIL

Labour law in Brazil was for long differentiated for agricultural and non-agricultural workers. The Labour Law Consolidation of 1943 did not apply to agricultural workers. The labour rights of agricultural workers were first protected only with the adoption of the Rural Workers Statute of 1963 and the creation of a specific social security institute (FUNRURAL). However, agricultural workers continued to enjoy lesser rights than their non-agricultural counterparts. In 1973, the 1963 Statute was repealed by Law 5889, which governs agricultural labour and states the applicability of the Labour Law Consolidation insofar as not inconsistent with it (Law 5889, art. 1). The dualistic labour protection was definitively repealed by the 1988 Constitution, which granted to both urban and rural workers labour rights such as protection against unfair dismissal, minimum wage, maximum working hours, weekly rest and annual paid leave, social security for unemployment and work-related injuries, safe and healthy working conditions, and collective bargaining and trade union rights (arts. 7 and 8).

As for women's access to employment, article 7(XXX) of the Constitution and article 373A of the Labour Law Consolidation (inserted by Law 9799 of 1999) prohibit discrimination on the basis of sex, pregnancy and marital status in recruitment, and envisage special measures to promote women's employment.

Marital authority provisions limiting women's access to employment have been repealed. The norms of the 1916 Civil Code requiring the authorization of the husband for women's employment (arts. 233(IV) and 242(VII)) were repealed by Law 4121 (1962). Article 446 of the Labour Law Consolidation, entitling the husband to rescind the employment contract of the wife if he deemed it necessary for household needs, was not applied after the 1962 Law and was formally repealed by Law 7855 (1989). However, marital authority remains in practice widely applied in rural areas (FAO, 1994).

A common discriminatory practice concerning women's access to employment is the requirement by employers of sterilisation certificates as a condition for recruitment (CEACR (111), 1993). Law 9029 of 1995 and article 373A(IV) of the Labour Law Consolidation (inserted by Law 9799 of 1999) prohibit employers from requiring sterilisation or pregnancy certifications or examinations as a condition for employment, and bar employers from conducting intimate examinations of employees. The phenomenon seems to have subsided since the mid-nineties.

As for treatment, article 7(XXX) of the Constitution and article 373A of the Labour Law Consolidation (inserted by Act 9799 of 1999) prohibit discrimination on the basis of sex, pregnancy and marital status in training, promotion and dismissal (except in case of incompatibility with the nature of the employment), as well as in remuneration. Work of the same function, productivity and technical sophistication must be paid with equal remuneration (Labour Law Consolidation, arts. 5 and 461). Training provided to employees by the government, employers and others must be available without sex discrimination (Labour Law Consolidation, art. 390B, inserted by Law 9799 of 1999).

However, in practice, a substantial gender pay gap exists (CEACR (111), 1994). Barsted (2002) reports that the average rural sector wage is R$ 257.97 for men and R$ 144.40 for women. Moreover, working conditions and wages differ considerably among women belonging to different racial groups; within each group, there is a considerable gender pay gap. Women make up most of the informal sector, and are therefore often not protected by formal legislation (Justiça Global, 2000). In this context of discrimination, there is an overall lack of complaints by rural women. In most cases, the victims of discrimination refuse to be identified for fear of reprisals and for lack of trust in public authorities (CEACR (111), 1994).

Pregnant workers have a right to a maternity leave of 120 days, without prejudice to employment and salary (art. 7(XVIII) of the Constitution). In 1996, the High Labour Court declared that maternity leave is a fundamental right which cannot be negotiated or alienated (CEACR (103), 1999). Pregnant workers cannot be dismissed from the date of the communication of the pregnancy to the employer to five months after childbirth; in case of dismissal, pregnant women have the right to be reinstated in their position (art. 10(II)(b) of the transitory constitutional provisions and Labour Law Consolidation, art. 392).

As for maternity benefits, rural occupations have been equated to urban occupations by the 1988 Constitution. This required implementing legislation, which was adopted only in 1994, after mobilizations of women workers (Guivant, 2001). Workers on maternity leave are now entitled to a benefit equivalent to the minimum wage, paid by the social security institution, provided that they prove to have worked for 12 (not necessarily continuous) months (Law 8213 (1991), art. 39, as amended by Law 8861 (1994)). Decree 4883 (1998) provided that maternity benefits were to be paid through social insurance only up to a limit, beyond which they were to be paid by employers; however, the Federal Supreme Court held this limit unconstitutional, stating that it was for the state to entirely pay maternity benefits (CRLP, 2000).

Pregnant workers carrying out work prejudicial to health may obtain transfer to another work (Labour Law Consolidation, art. 392, as amended in 1999). Nursing women have the right to two half-an-hour nursing breaks per day for children up to the age of six months. Article 7(XXV) of the Constitution grants rural workers the right to free day-care for children from birth to six years. Article 389(1) of the Labour Law Consolidation requires employers employing 30 or more women over 16 years to provide a crèche for children. However, Legislative Decree 229 (1967) allows employers to provide crèche reimbursements instead. Agricultural undertakings employing more than 50 families of workers must provide free primary schools for the children of the farm workers (Law 5889 (1973), art. 16).

While social benefits (maternity leave, retirement pension, etc.) apply to all rural workers, their actual enjoyment is conditional upon presentation of documentation (identity card, fiscal registration number - CPF, work card, etc.). Due to monetary and transaction costs, few rural women have these documents. In Rio Grande do Sul, for instance, 30 percent of the women rural workers did not even have the identity card (Guivant, 2001).

As for sanctions, Law 9459 of 1997 amends previous legislation, providing for tougher criminal sanctions for violations of the non-discrimination principle. However, as of 1999 no cases had been brought under this law (CEACR (111), 1999).

Relevant provisions have also been adopted at state and municipal level. For instance, the municipality of São Paulo adopted Law 11081 (1991) and Decree 30497 (1991), empowering the municipality to impose sanctions on employers requiring pregnancy tests, gynaecological examinations or sterilisation certificates to obtain or maintain jobs.

Box 1
Women's Access to Employment in Guatemala: The Morales De Sierra Case

In Guatemala, the Civil Code allowed married women to undertake an occupation only insofar as consistent with their role as housewives (art. 113), and allowed the husband to oppose the employment of his wife, provided that he had sufficient earnings to provide for the maintenance of the household and he had justified reasons (art. 114).A constitutionality challenge of these discriminatory norms was rejected by the Constitutional Court on the basis inter alia of the need to ensure legal certainty and to protect the children (Case 84-92 of 1992). Another constitutionality challenge was brought before the Constitutional Court by the Attorney-General of Guatemala in 1996. In 1995, a woman filed a complaint with the Inter-American Commission on Human Rights, challenging articles 113 and 114 as well as other provision of the Civil Code concerning the administration of family property.

In 1998, deciding on a preliminary controversy concerning locus standi (the woman had not suffered from the application of the challenged norms herself) and exhaustion of domestic remedies, the Commission admitted the complaint (Maria Eugenia Morales de Sierrav. Guatemala, Inter-American Commission on Human Rights, Case 11625, Report No. 28/98, 6 March 1998). While the proceeding was pending, most of the challenged norms (including articles 113 and 114) were repealed by Decrees 80 (1998) and 27 (1999), reforming the Civil Code. In addition, Decree 7 of 1999 (Ley de Dignificación y Promoción Integral de la Mujer) was adopted, guaranteeing women's right to freely choose their employment and prohibiting discrimination on the basis of marital status (art. 12). In 2001, the Inter-American Commission issued a report on the merits. The Commission clarified that differences in treatment do not necessarily amount to discrimination, where they are based on "reasonable and objective criteria". The Commission held however that the challenged provisions could not be justified, and violated articles 11, 17(4) 24 of the ACHR. The Commission recognized the important progress made with the 1998 and 1999 reforms, and called the state to fully comply with its international human rights obligations (Maria Eugenia Morales de Sierrav. Guatemala, Inter-American Commission on Human Rights, Case 11625, Report No. 4/01, 19 January 2001).

SUB-SAHARAN AFRICA

REGIONAL OVERVIEW

The ACHPR recognises the right of "every individual" to work under equitable and satisfactory conditions and to receive equal pay for equal work (art. 15). On the other hand, the Charter is silent on some other aspects concerning labour rights, particularly with regard to trade union rights. Non-discrimination in training and equal opportunities to work (including women's freedom to choose their occupation, equality in access to employment, and equal remuneration for jobs of equal value) are affirmed in the Draft Protocol on the Rights of Women in Africa (not yet adopted; arts. 12 and 13).

In several countries sex discrimination in employment is prohibited by constitutional norms (e.g. art. 17(3)(e) of the 1999 Constitution of Nigeria, stating the equal pay principle among the Fundamental Objectives and Directive Principles of State Policy) and/or by ordinary legislation (e.g. Labour Code of Ivory Coast 1995, art. 4, prohibiting sex discrimination in recruitment, promotion, remuneration, vocational training, labour division, social benefits and termination of the labour contract; art. 14(1)(b) of the Ethiopian Labour Proclamation of 1993, prohibiting sex discrimination in remuneration). However, labour legislation often excludes the agricultural sector from its scope of application. For instance, Nigeria's National Minimum Wage Act of 1981, as amended, excludes workers in farms employing fewer than 50 persons, part-time workers, workers paid on piece-rate and seasonal agricultural workers (CEACR (100), 2000).

Maternity leave ranges from 60 days (Mozambique, Guinea Bissau, Eritrea) to 14 weeks (Cameroon, Central African Republic, Chad, Gabon, Madagascar, Togo). Cash benefits range from 25 percent of the wage (Botswana) to 100 percent (Congo, Mauritania, Mauritius, Zambia), with a substantial number of countries granting 50 percent of the wage (Chad, Burundi, Ghana, Nigeria). No benefit is paid in some countries (Lesotho). In some countries cash benefits are funded by social security institutions (Namibia, Senegal), in others they are paid by the employer (Ethiopia, Ghana, Nigeria) (United Nations, 2000, updated to 1998). The requirements to qualify for these benefits may be very demanding; for instance, the Zambian Employment Act of 1965 requires at least two years of continuous service with the employer (secs. 15A and B, inserted by Law 18 of 1982[38]). In some countries, legislation prohibits dismissal during pregnancy (e.g. the Labour Code of Ivory Coast; art. 23(3), and the 1975 Labour Act of Nigeria, sec.53(4)).

However, laws protecting women's labour rights are often not implemented. Rural women are often unaware of their legal rights. In addition, women make up a considerable portion of the agricultural labour force employed in the informal sector (which accounts for substantial GDP shares throughout sub-Saharan Africa; see e.g. MacGaffey, 1991), where labour legislation is not applied.

Field studies from plantations (e.g. Mbilinyi, 1995, on a sugar cane plantation in Tanzania; Bob, 1996, on a coffee plantation in South Africa) found evidence of widespread discrimination in access to employment (with women concentrated in low-pay and subordinate manual jobs in the fields and men in higher positions, particularly as supervisors and headmen), wage differentials (with higher wages for typically men's positions, e.g. sugar cane cutters, than for women's positions, e.g. weeders), sexual harassment (most often by headmen), discrimination in access to training and vocational courses, discrimination in benefits allocation (e.g. where housing is provided, unmarried workers are given housing units suitable for men without dependants but extremely small for female-headed households) and discrimination within trade unions (as for participation, leadership positions, etc.). Discriminatory provisions may also be contained in collective agreements concerning plantation workers.

Customary law also affects the labour rights of women agricultural workers. Generally speaking, there is a gender division of labour, whereby men mainly cultivate cash crops, while women cultivate food crops or locally traded crops. However, under many customary legal systems, women must provide their labour for certain tasks in their husbands' fields (e.g. weeding). This work, provided within the household, is unpaid and unprotected (Lastarria-Cornhiel, 1997).

KENYA

Agricultural labour is covered by the Employment Act of 1976 (Cap. 226). This Act fails to address gender issues. No reference is made to the principle of non-discrimination. No provision exists on sexual harassment. Women's night work is prohibited in industrial undertakings (with exceptions; secs. 28 and 29); the Ministry of Labour may prohibit or subject to conditions women's work in "any specified trade or occupation" (sec. 56(1)(j)).

Minimum wage legislation (Regulation of Wages and Conditions of Employment Act, Cap 229) provides for the establishment of sector-specific wages councils by the Ministry for Labour, with the task of recommending wage determination or regulation for specific trades or occupations. Wages councils have been established for several sectors (e.g. textiles, domestic servants, tourism, etc.), including agriculture. Neither the Act nor the Regulation of Wages (Agricultural Industry Wages Council Establishment) Order, as amended, which establishes the wages council for agriculture, explicitly refer to the equal pay principle.

The Regulation of Wages (Agricultural Industry Wages Council Establishment) Order also determines maximum working hours. For male workers, the limit is 42 hours over six days per week (72 hours over seven days for some occupations, e.g. herdsmen); for female workers, the limit is 36 hours over six days per week (sec. 5).

The number of women employed in the formal sector has increased, mainly because of women's improved access to education. However, a gender division of labour (with higher positions being usually reserved to men) remains, due to cultural attitudes rather than to formal legislation. Moreover, women are mostly concentrated in the informal sector. Although women's wages relative to men's have increased in the last decades, a considerable gender pay gap nevertheless remains. In rural areas, women are largely unaware of their legal rights (Gopal and Salim, 1998).

Maternity protection was originally established with a 1975 Presidential Directive, envisaging a two-month paid maternity leave. The Employment Act of 1976 provides for a two-month fully-paid maternity leave at the expense of the employer (sec. 7(2)). The leave period is considerably shorter than that envisaged by international labour conventions. Moreover, women taking the maternity leave lose their one-month annual leave for the relevant year (sec. 7(2)). Furthermore, obliging employers to pay for maternity benefits raises the cost of women's labour and therefore discriminates against them (House-Midamba, 1993). The Act does not explicitly prohibit the dismissal of pregnant women.

Under the pension law, widows (though not widowers) lose their work pension upon remarriage (CRLP, 1997).

BURKINA FASO

The Constitution of Burkina Faso states that everybody has an equal right to work, and prohibits sex discrimination in employment and remuneration (art. 19).

The Labour Code, as revised in 1992 (Act 11 of 1992[41]), applies to all employment contracts (art. 1(1) and (2)). The Code states the principle of non-discrimination on the basis of sex (sec. 1(3)). However, it does not envisage sanctions for violations of this principle.

The minimum age for work is 14 years, without distinction between men and women (art. 87). Article 104 of the Labour Code states the principle of equal pay for equal working conditions, professional qualifications and output. No reference is made to work of equal value.

Maternity leave is of 14 weeks (six of which before and eight after childbirth), with maternity benefits equivalent to 100 percent of the wage paid jointly by social security institutions and employers (art. 84). Nursing mothers have two daily breaks for up to 15 months (art. 85).

Article 82 of the Labour Code prohibits assigning workers to tasks that may endanger their reproductive capacity or, in the case of pregnant workers, the health of the workers or of their child.

Notwithstanding this legislation, women's participation in formal employment is very low compared to men's (just over 12 percent over the period 1986-1992), without major trends towards improvement (CEACR (111), 1995). Sex/gender discrimination in recruitment, allocation of responsibilities, and remuneration has been reported (CEDAW, 2000).

Customary law contains labour obligations for women. In Comoé Province, young wives have the duty to provide labour for their husbands' fields, in addition to cultivating their own fields. The extent of this duty varies across ethnic groups, with particularly extensive labour obligations among the Turka and the Gouin. Women are liberated from these obligations usually in their mid-forties, when their children are old enough to provide labour (van Koppen, 1998).

SOUTH AFRICA

Until recently, women farm workers had little protection under South African legislation. The Basic Conditions of Employment Act of 1983 was extended to agricultural workers in 1993. The Labour Relations Act of 1995, the Basic Conditions of Employment Act of 1997 and the Employment Equity Act of 1998 have substantially improved the position of women farm workers.

For a long period, the legal capacity of married women to sign employment contracts was restricted by family law. Under the Black Administration Act of 1927, women married within a customary marriage could not sign contracts without the assistance of their husband (sec. 11(3)). This norm was repealed by the Recognition of Customary Marriages Act of 1998, granting wives full legal capacity to sign contracts (sec. 6).

The Employment Equity Act of 1998 prohibits direct and indirect unfair discrimination on grounds of gender, sex, pregnancy, marital status, and family responsibility in both access (recruitment) and treatment (job classification, remuneration, employment benefits, employment terms and conditions, promotion and dismissal) (secs. 6 and 1). Where discrimination is alleged, the burden of proof on its fairness is placed on the employer (sec. 11). The principle of non-discrimination on grounds of sex and pregnancy has also been affirmed in the case law, particularly in McInnes v. Technikon Natal (Labour Court, D322/98, March 2000) and Woolworths (Pty) Ltd. v. Whitehead (Labour Appeal Court, CA06/99, 3 April 2000).

Besides prohibiting discrimination, the Employment Equity Act also provides for affirmative action with regard to "designated employers" (i.e. those employing 50 or more workers, or less than 50 but with a determined annual turnover) (sec. 1 and Chap. III). Other employers may voluntarily comply with the affirmative action rules (sec. 14). Affirmative action measures must be taken by designated employers in consultation, and possibly agreement, with employees (sec. 16). Employers are obliged to develop "employment equity plans" must state objectives, numerical targets, strategies, measures, timetable and monitoring and evaluating procedures (sec. 20).

Affirmative action measures may include preferential treatment and numerical goals (but not quotas) (sec. 15).

Moreover, "designated employers" are subject to special rules on the application of the equal remuneration principle. In particular, they are to report remuneration and benefits to the Employment Conditions Commission; where income differentials are disproportionate, they must adopt measures, including collective bargaining, application of the standards set by the Employment Conditions Commission, and promotion of training (sec. 27).

Farm workers employed for more than four weeks have the right to a four-week contract termination notice (the same as workers employed for more than one year in non-agricultural sectors) (sec. 39(1)(c) of the Basic Conditions of Employment Act).

In practice, the employment of women farm workers is often tied to their husband's employment. Indeed, there are reports that married women farm workers are denied contracts in their own names, and work on the basis of contracts signed by their husbands (Human Rights Watch, 2001).

A field study from the Waterval Coffee Plantation in Lebowa found no formal discrimination and yet substantial gender differences as to the nature of the job and the remuneration, with women concentrated in low-pay, seasonal/temporary jobs. Women constituted 22 percent of the permanent workers (2 out of 9), 66 percent of the semi-permanent workers (36 out of 54), and 92 percent of the seasonal and temporary workers (275 out of 300). This gender stratification of labour had an effect on the wage structure. Permanent workers (mainly men) were paid R18.00 per day, while seasonal and temporary workers were paid on piece-rate at R2.80 per crate (with an average of two crates per day). Tenure security was also unequally distributed, as positions dominated by men (permanent jobs) were more secure than those mainly occupied by women (seasonal and temporary workers); temporary and seasonal workers had no security of being hired the following year. Moreover, women bore a very heavy workload, cumulating work in the plantation with domestic responsibilities (Bob, 1996).

Housing benefits granted to plantation workers are usually given to (male) permanent employees, while single women (usually temporary) workers tend to be excluded (Human Rights Watch, 2001). In this regard, case law has been developed under the Extension of Security of Tenure Act (ESTA), which protects from eviction persons occupying land with the consent of the landowner on the date of the Act (including farm workers). In Landbou Navorsingsraad v. Klaasen (LCC 83R/01, 29 October 2001), the Land Claims Court held that the wife of an "occupier" protected under the Act (a farm worker) is not entitled to an eviction notice under section 9(2)(d) of ESTA unless the landowner explicitly consented to her residence on the land.

In Conradie v. Hanekom and Another (1999 (4) SA 491 LCC), the Land Claims Court set aside an eviction order against two farm workers, husband and wife, employed on the same farm. Having dismissed the husband, the landowner had sought to evict both. The court held that the wife, as employee, had a right not to be evicted under ESTA, and her eviction order was set aside. Moreover, she had a right to family life under section 6(2)(d) of ESTA; therefore, her husband (who after his dismissal was no longer a protected "occupier") had a right to reside in the land as a family member of an "occupier".

Another interesting case, concerning inter alia the protection from eviction for family members of farm workers, is Lombaard NO v. Motsumi and Others (LCC 52R/01, 17 May 2001). In this case, the applicant sought to evict farm occupiers and their household members after termination of their employment contracts. While the magistrate court granted an eviction order, the Land Claims Court set it aside on the ground that while employment was terminated, many employees and household members were occupiers under ESTA on other grounds (e.g. many of the household members were born and had lived in the farm for all their life).

As for maternity protection, the Basic Conditions of Employment Act of 1997 provides that a female employee is entitled to at least four months maternity leave, of which six weeks must be taken following childbirth. Pregnant or nursing women cannot be required to perform work hazardous to her health or to the health of the child; male and female employees have a annual three-day family responsibility paid leave for childbirth or child illness (secs. 25-27).

Maternity benefits are paid by the Unemployment Insurance Fund under the Unemployment Insurance Act. Under this Act, maternity benefits are 45 percent of a worker's normal weekly earnings for a period not exceeding 26 weeks, provided that the worker has been employed for at least 13 weeks during the 52 weeks before childbirth (secs. 34 and 37). However, this Act does not apply to seasonal farm workers, i.e. workers employed for less than a continuous period of 4 months. Moreover, the Fund manages unemployment, illness and maternity benefits, and there is an overall limit on the benefits that a worker can receive (one week's benefits for each six weeks' employment); therefore, women on maternity leave use up their rights to unemployment benefits. At the time of writing, amendments to reform this system have been proposed.

The Labour Relations Act of 1995 prohibits unfair dismissal, which includes both the failure of an employer to allow the return of a woman worker after maternity leave on the one hand, and the renewal of a temporary employment contract on less favourable terms after maternity on the other; dismissal for pregnancy (as well as intended pregnancy and any reason relating to pregnancy) or on directly or indirectly discriminatory grounds is automatically unfair (secs. 186 and 187).

The Promotion of Equality and Prevention of Unfair Discrimination Act of 2000 applies where the Employment Equity Act does not apply (sec. 5(3)). It prohibits unfair discrimination against women by the state and any persons, for instance in women's access to social security (secs. 6 and 8(g)).

As for social security laws, old age pension constitutes a major source of income for the rural poor. Although women constitute the majority of the eligible population (because of women's lower retirement age - 60 instead of 65 - and because of their longer life expectancy), fewer women than men benefit from the pension programme. One of the explanations for this is that many rural women may lack identity cards, which are required for pension eligibility (Baden *et al.*, 1999).

compared e.g. to 55 percent in the textile industry) (Belarbi *et al.*, 1997).

Article 64 of the Labour Code, concerning maternity protection, applies to all enterprises, thus including agricultural ones, with the only exception of family enterprises. It grants pregnant workers a paid maternity leave of 30 days. Maternity leave may be extended for 15-day periods with the supply of a medical certificate. Nursing women have two half-an-hour breaks per day to breastfeed. Establishments employing more than 50 women must provide a room for breast-feeding.

As for social security, the general social security regime, originally not covering agricultural labour, was extended to agricultural workers in 1970, provided that the workers have been employed for more than six months with the same employer (Belarbi et al., 1997). Given their concentration in seasonal and temporary positions, women workers may find it difficult to meet this requirement.

ASIA

REGIONAL OVERVIEW

The principle of non-discrimination on the basis of sex/ gender is affirmed in several countries. For instance, the 1998 Labour Protection Act of Thailand mandates employers to treat male and female employees equally with regard to employment, "except where the nature or conditions of the work does or do not allow the employer to do so" (sec. 15). On the other hand, women's access to agricultural employment is often limited by the extension of night work bans to the agricultural sector (e.g. the 1955 Employment Act of Malaysia, sec. 34(1)) and by provisions requiring the authorization of the husband for women to sign employment contracts (e.g. Civil Code of Indonesia, art. 1601(f)).

Existing studies suggest the existence of rigid gender occupational segregation. In Bangladesh, for instance, women's employment in the agricultural sector increased substantially in recent years, but remains concentrated in seasonal occupations and remunerated with wages lower than those for men. Labour legislation is poorly enforced, and trade unions rarely protect the interests of women workers as such (Baden *et al.*, 1994).

Maternity leave ranges from 52 days (Nepal) to 12 weeks (Bangladesh, Pakistan), but is generally short (e.g. 60 days in Malaysia; 90 days in Cambodia, China and Laos). Cash benefits range from 50 percent of the wage (Cambodia) to 100 percent (most countries: e.g. China, Indonesia, Malaysia, Nepal). In most countries maternity benefits are paid by the employer (Bangladesh, Indonesia, Nepal, Malaysia, Sri Lanka), although some exceptions exist (benefits are funded through social security in the Philippines and in Viet Nam) (United Nations, 2000, updated to 1998). Some laws specifically prohibit dismissal on grounds of pregnancy (e.g. Thai Labour Protection Act; sec. 43, Malaysian Employment Act, art. 40(3)).

INDIA

Article 39 of the Indian Constitution directs the state to ensure that "citizens, men and women equally, have the right to adequate means of livelihood"; that "there is equal pay for equal work for both men and women"; and that "the health and strength of workers, men and women, [...] are not abused and that citizens are not forced by economic necessity to enter avocations unsuited to their age or strength". On the other hand, article 16 of the Constitution, stating the principle of equality in employment, applies to public employment only.

With regard to access to employment, sex discrimination is prohibited by the Equal Remuneration Act of 1976 (sec. 5, as amended in 1987). In practice, however, in many rural areas women's access to employment is restricted by cultural factors such as female seclusion (purdah) and the perception of women's abstention from work as an indicator of the social status and success of the husband. Moreover, a gender division of labour remains widespread, with women concentrated in "feminine" jobs, particularly low-skill, low-pay agricultural work (e.g. weeders) (Jha et al., 1998; GoI, n.d.).

With regard to treatment, the Equal Remuneration Act of 1976, as amended, prohibits discrimination in employment conditions (including promotion, training and transfer) (sec. 5, as amended in 1987). On the other hand, protective legislation prohibits women's night work in a number of sectors. As for agriculture, the Plantations Labour Act of 1951 prohibits the employment of women between 19 hours and 6 hours (except for midwives and nurses) unless there is a permission from the state government (sec. 25).

The Equal Remuneration Act states the principle of equal remuneration for the "same work or work of a similar nature" (though not for work of equal value; sec. 4). In complying with this requirement, employers cannot reduce wages; therefore, in case of existing sex discrimination, the higher wage is payable to workers of both sexes (sec. 4). The equal remuneration principle is also guaranteed in the case law (Mackinnon Mackenzie & Co. v. Audrey D'Costa, 1987 2 SCC 469).

In practice, substantial gender pay gaps exist: women's wages are lower than men's in all states of the federation (on average, 30 percent lower); there is no institutional machinery for the implementation of minimum wage legislation in the agricultural sector (United Nations, 1997; Menon-Sen and Kumar, 2001). According to the Centre of Indian Trade Unions (CITU), the 1976 Act is mainly applied to public sector industries, while gender pay gaps persist in other industries, including agriculture, where employers avoid the application of the minimum wage legislation by paying workers on a piece-rate basis; in these sectors, female workers are paid considerably lower wages than male workers (CEACR (100), 1998).

Indian law contains no specific provision on sexual harassment in the workplace. However, the Supreme Court developed guidelines in Vishaka v. State of Rajasthan and Others (AIR 1997 SC 3011). The guidelines are to be applied in all workplaces, and build on the Indian Constitution, on the CEDAW and on General Recommendation No. 19 of the CEDAW Committee (on violence against women).

Maternity leave is governed by the Maternity Benefit Act of 1961, which applies to plantations and to other establishments with more than ten employees (sec. 2). Pregnant workers have a right to 12-week paid maternity leave (secs. 4 and 6(2)). A six-week leave is granted in case of miscarriage or termination of pregnancy (sec. 9). An additional one-month leave is provided in case of illness arising out of pregnancy, delivery, miscarriage or termination of pregnancy (sec. 10). Maternity benefits are equivalent to the average daily wage of the woman worker (sec. 5). Discharge or dismissal of a woman on maternity leave, as well as the varying of her working conditions at her disadvantage, are prohibited (sec. 12).

Pregnant workers have the right not to perform arduous work, or work which involves long hours of standing or which is likely to interfere with the pregnancy, the normal development of the foetus, adversely affect health or cause a miscarriage (Maternity Benefit Act, sec. 4(3)). No deductions from wages can be made because of the changed nature of the work performed (sec. 13 of the same Act). Nursing women have a right to two nursing breaks per day until the child attains the age of fifteen months, without deductions from the wage (secs. 11 and 13). Under the Plantations Labour Act of 1951, employers with more than 50 women workers (or with women workers having a number of children under six years old of 20 or more) must provide crèche facilities (sec. 12).

The laws on social security (Employees' Provident Fund and Miscellaneous Provisions Act 1952 and the Payment of Gratuity Act 1972) apply equally to men and women.

THE PHILIPPINES

Agricultural labour is governed by the Labour Code of 1974, as amended (art. 6)[44]. Article 3 of the Code declares that the state is to ensure equal work opportunities regardless of sex. However, sex discrimination with regard to recruitment is not explicitly prohibited.

Article 136 of the Code declares that it is unlawful for an employer "to require as a condition of employment or continuation of employment that a woman employee shall not get married, or to stipulate expressly or tacitly that upon getting married, a woman employee shall be deemed resigned or separated, or to actually dismiss, discharge, discriminate or otherwise prejudice a woman employee merely by reason of her marriage".

As for treatment, sex discrimination with respect to terms and conditions of employment is prohibited. Acts of discrimination include "payment of a lesser compensation, including wage, salary or other form of remuneration and fringe benefits, to a female employees as against a male employee, for work of equal value", and "favouring a male employee over a female employee with respect to promotion, training opportunities, study and scholarship grants solely on account of their sexes". A wilful violation of this provision entails criminal responsibility (Labour Code, art. 135). Discrimination against indigenous women in the areas of employment and training is prohibited under sections 21, 23 and 25 of the Indigenous Peoples Rights Act of 1997.

Women's night work in agricultural undertakings is prohibited unless women are granted a period of rest of at least nine consecutive hours (Labour Code, art. 130(c)). The Secretary of Labour and Employment is to set standards to ensure the safety and health of women employees and to enact regulations requiring employers to provide facilities for women workers (separate toilets, nurseries, etc.) (Labour Code, art. 132). Sexual harassment in the workplace is prohibited by the Anti-Sexual Harassment Act of 1995.

Notwithstanding these provisions, gender labour differentiation remains. Women are concentrated in "feminine" occupations. As for agriculture, while men are considered "farmers" (i.e. farm heads), women are usually referred to as "farm workers" (Roces, 2000). Gender pay gaps also remain. In 1990, women's average wage was about 40 percent that of men. In this regard, agriculture is a particularly difficult sector, as in 1989 women's average income was about 10 percent that of men (United Nations, 1995). A considerable number of female agricultural workers (about 50 percent) are unpaid (Roces, 2000).

Pregnant workers have the right to a six-week fully paid maternity leave (two weeks before and four weeks after childbirth), extendable without remuneration in case of illness arising out of pregnancy, delivery, abortion or miscarriage. However, maternity benefits are granted only for the first four deliveries (Labour Code, art. 133). Maternity benefits are to be paid by the employer, although under the Maternity Benefits Act of 1992 full maternity benefits are paid by the Social Security System for 60 days for women workers meeting specified requirements. Employers cannot discharge pregnant workers on account of their pregnancy or while on maternity leave, nor discharge or refuse their admission upon returning to their work for fear that they may be pregnant again (Labour Code, art. 137(a)(2) and (3)).

As for social security, the Social Security System covers all employees, without making distinctions on the basis of sex or gender (Labour Code, art. 168, and the Social Security Act of 1997, sec. 9(a)). Housewives may be covered by the Social Security System on a voluntary basis (Social Security Act, sec. 9(b)).

THE PACIFIC REGION

REGIONAL OVERVIEW

Of all the countries of the region, only Australia, New Zealand and Papua New Guinea have ratified ILO Conventions 100 and 111. In several countries, labour legislation does not explicitly prohibit discrimination on the basis of sex/gender (e.g. Fiji, Samoa, Tonga). Papua New Guinea is an exception, as article 48 of the Constitution states that "every person" has a right to freedom of choice of employment and section 97 of the Employment Act prohibits discrimination on the basis of sex. The Employment Act of Vanuatu prohibits sex discrimination but only with regard to remuneration (sec. 8).

Maternity leave is six weeks in Papua New Guinea and 12 weeks in the Solomon Islands. Maternity benefits are generally not paid (e.g. Papua New Guinea, Australia and New Zealand) or very low (25 percent of the wage in Solomon Islands); when benefit payment is envisaged, it is usually paid by the employer (e.g. Solomon Islands) (United Nations, 2000, updated to 1998). Protection from dismissal during pregnancy is envisaged in some countries, but the covered period is usually very limited (a few days beyond maternity leave in Fiji; three weeks beyond maternity leave in Vanuatu).

FIJI

Fiji has not ratified any of the relevant ILO Conventions. With a Cabinet Decision of 7 December 2001, the government decided to ratify five ILO Conventions, including Conventions 100 and 111. However, at the moment of writing no step in this direction had been taken yet.

Agricultural labour is governed by the Employment Act (Cap 92), which is currently being reviewed by the government (GoF, 1999). However, service contracts for the harvesting of sugar cane, which is Fiji's major cash crop, are excluded from the scope of the Act by the Employment (Application) Order (sec. 3 and second schedule).

The Employment Act does not specifically state the principle of non-discrimination on the basis of sex or gender. It contains a ban on women's night work, which does not apply to the agricultural sector (sec. 65). No provision specifically deals with sexual harassment in the workplace.

Maternity protection is very limited. Maternity leave is of 84 days, to be divided in two periods of 42 days each before and after childbirth (sec. 74(1)). Protection from dismissal extends to three months, i.e. just a few days beyond the maternity leave period (sec. 79(1)). Maternity benefits are determined through a flat rate amount, which is very low and in most cases lower than full pay (US$1.50 per day; sec. 74(1)). Maternity benefits are paid integrally by the employer (sec. 74). Lack of pregnancy notification to the employer within the terms, and defect or inaccuracy of the notification where the defect or inaccuracy causes prejudice to the employer, entail loss of maternity benefits (sec. 77).

Under the Married Women's Property Act of 1892 (Cap 37), a married woman is entitled to hold and dispose of her separate property, including "any wages, earnings, money and property gained or acquired by her in any employment, trade or occupation in which she is engaged or carries on separately from her husband" (sec. 4).

Although there has been an increase in women's participation in formal employment, women remain confined to low paying jobs, and concentrated in the informal sector (with little or no security), principally in subsistence agriculture. There is little awareness of labour rights among women, including on maternity leave legislation. Discriminatory practices include unequal remuneration, unequal training and career opportunities, and sexual harassment (GoF, 1999).

EUROPE

REGIONAL OVERVIEW

While the ECHR is silent on socio-economic rights, the European Social Charter, as revised, recognises the right of all workers to equal opportunities and equal treatment in employment and occupation, without discrimination on the ground of sex, and states the principle of equal remuneration for work of equal value. The Charter also guarantees the right to paid maternity leave for at least 12 weeks, funded by social security institutions or by public funds.

For countries members of the European Union, EU legislation and case law on non-discrimination in employment applies. Article 2 of the EC Treaty includes gender equality among the objectives of the European Community. Under article 13 of the EC Treaty, the Council of Ministers may take action to combat sex discrimination. Article 141(1) (formerly 119) of the EC Treaty (as amended) states the principle of equal remuneration for work of equal value. Article 141(4) provides for affirmative action. Equality of treatment for men and women in access to employment and vocational training is stated in Directive 76/207 of 1976, while the principle of equal pay for equal work is implemented by Directive 75/117 of 1975. Under Directive 97/80 of 1997, when persons alleging discrimination violating Directives 117 and 207 establish prima facie discrimination (i.e. "facts from which it may be presumed that there has been direct or indirect discrimination"), the burden of proof is on the employer to prove that no violation occurred. A three-month parental leave is granted to both parents to tend children up to eight years old (Directive 96/34 of 1996). A vast case law on gender equality in the workplace exists within EU law.

The principle of non-discrimination on the basis of sex/ gender is stated in all Western legal systems. Some countries also provide for special measures to promote women's employment. However, there are reports that in the rural areas of some EU states, women's unemployment rates are higher than men's, due to traditional attitudes on the role of men and women and to the shortage of transport and care facilities (Braithwaite, 1996).

As for Central and Eastern European countries, the principle of equal treatment for men and women is affirmed e.g. in Croatia (Labour Act 1995, art. 82) and in Romania (art. 38(4) of the Constitution and Labour Code, arts. 14 and 151(1)). A considerable gender pay gap has however been documented for several countries (UNICEF, 1999).

As for maternity leave, most countries comply with the international standard of 12 weeks. Cash benefits range from 75 percent of the wage (Greece) to 100 percent (most countries: e.g. Germany, Poland, Russia); in some cases, benefits vary during the leave period (e.g. 82 percent of the wage for 30 days and 75 percent thereafter in Belgium). In most European countries maternity benefits are paid through social security (Belarus, France, Germany, Hungary, Romania, United Kingdom) (United Nations, 2000, updated to 1998). Pregnant women are usually protected from discrimination or dismissal (e.g. in Croatia, Labour Code 1995, art. 55).

ITALY

In Italy, agricultural labour is governed by general labour law, although some aspects are governed by specific norms, such as recruitment procedures, work reinstatement following unjust dismissal, employment contract duration and social security. Moreover, the above-mentioned norms of EU law apply; Law 52 of 1996 provided for the implementation of EC Directives relating to equal opportunities (art. 18).

Direct and indirect discrimination on the basis of sex, marital or family status or pregnancy is prohibited with regard to both access and treatment. Where the person alleging discrimination proves facts establishing prima facie discrimination, the burden is on the employer to prove the absence of discrimination. The equal remuneration principle is stated with regard to both equal work and work of equal value (Article 37 of the Constitution; Law 903 of 1977, arts. 1-3; Law 125 of 1991, art. 4(6), as amended; Legislative Decree 151 of 2001, art. 3).

Contractual clauses envisaging termination of the employment of women workers in case of marriage, as well as actual dismissal for marriage and worker's resignation within one year from marriage (unless confirmed before the labour office), are null and void, and the worker has the right to be reinstated in her position. Dismissals within one year from marriage are presumed to be motivated on grounds of marriage unless the employer proves otherwise (Law 7 of 1963, art. 1).

Affirmative action measures to promote women's employment and de facto equality of opportunity can be undertaken by employers, with public funding (Law 125 of 1991, as amended by Legislative Decree 196 of 2000). Projects implemented so far concerned mainly access to male-dominated sectors and changes in the work organization and time. The implementation of the law is monitored by the National Committee for Equal Opportunities established within the Ministry for Labour. The major problems encountered in the implementation of the law include limited resources and cumbersome administrative procedures (Golt, 1999).

Maternity leave is of two months before childbirth (three months for dangerous and heavy jobs) and three months after childbirth, extendable for periods of two months in cases of illness arising out of pregnancy or childbirth. Maternity benefits are equivalent to 80 percent of the remuneration. For women workers in sharecropping undertakings (mezzadria and colonia), maternity benefits are 80 percent of the average daily income, as determined by the Ministry of Labour every two years (Law 1204 of 1971 and Legislative Decree 151 of 2001). Benefits are paid by the social security institution (INPS), funded through payroll taxes (increasing the cost of labour, though without distinction on the basis of sex), although there is a trend toward the gradual transfer of the funding for maternity benefits to the general taxation system (Golt, 1999).

Women workers cannot be dismissed from the beginning of pregnancy until the child reaches the age of one year; in case of dismissal, they have the right to be reinstated. Seasonal workers have the right to priority in seasonal recruitment. Women cannot be required to perform dangerous, tiring or unhealthy jobs during pregnancy and until seven months after childbirth, without wage deductions. Two nursing breaks per day are allowed (Laws 1204 of 1971 and 53 of 2000; Legislative Decree 151 of 2001). The safety and hygiene of working conditions for pregnant and nursing women are protected by Legislative Decree 645 of 1996.

In case of death, grave inability or abandonment of the mother, or in case of child custody to the father, the latter has the right to a paternity leave on the same terms of the maternity leave (Law 903, art. 6bis, and Legislative Decree 151 (2001), arts.

28 and 29). Parental leave of up to 10 months, until the child reaches the age of 8, is granted to both parents (Law 53 (2000), art. 3, and Legislative Decree 151 of 2001, art. 32).

As for social security, family benefits and pension increases for family dependants can be paid to working or retired women. Payment of social security benefits to the surviving spouse of the insured worker applies equally to men and women (Law 903 of 1977, arts. 9-12).

Italy, especially in the South, has a large informal economy compared to other developed countries. In the informal sector, labour legislation, including its provisions on gender equality, is not applied. Although gender-disaggregated data on this sector are scarce, women constitute a substantial portion of the informal labour force, including in the agricultural sector. Legislative efforts to promote the regularisation of the informal sector have been made (e.g. Laws 608 of 1996 and 196 of 1997, envisaging incentives for informal sector enterprises to register) (GoIt, 1999). Even in the formal sector of the economy, however, compliance problems may arise. A gender division of occupations and a gender pay gap still exist in several sectors, with substantial cross-sectoral variation (CEACR (100), 1998).

CONCLUSION

In the light of the analysis of the legislation of the covered countries, it is possible to highlight some key issues affecting the labour rights of women agricultural workers. First, women's access to employment may be restricted by family law norms requiring the authorization of the husband (e.g. in some Mexican states). In several countries, these norms have been successfully challenged by women at national and international level, and legislative reforms have been adopted to repeal them (e.g. in Guatemala). However, even where these norms are repealed, there are reports that marital authority continues to be applied in practice, especially in rural areas (e.g. Brazil).

Second, while labour legislation in most of the covered countries explicitly prohibits sex discrimination, in some cases labour laws are silent on this issue (e.g. Kenya and Fiji). In yet other cases, while the non-discrimination principle is formally stated, no sanction is envisaged for violations (Burkina Faso). In most of the covered countries (with some exceptions; e.g. Italy), sex discrimination in remuneration is prohibited only with reference to "equal work", not with reference to the internationally recognized criterion of "work of equal value". In all these cases, discriminatory practices on the workplace are de jure or de facto allowed. Affirmative-action measures are envisaged only in some of the covered countries (e.g. Brazil, South Africa and Italy).

Third, women's access to agricultural work may be hindered by "protective" legislation prohibiting women's night work in the agricultural sector, while international conventions prohibit it only with regard to some industrial occupations (e.g. India).

Fourth, only some countries have adopted legislation addressing sexual harassment on the workplace (the Philippines). In India, the legislative lacuna was filled by guidelines adopted by the judiciary. In other countries, sexual harassment is left unaddressed (Kenya and Fiji). Field studies document that this is a major problem affecting women working in plantations.

Fifth, women enjoy special protection in case of maternity in all the countries examined. However, requirements for the application of this protection (e.g. in terms of duration of previous employment and of documentation to be produced) may be very demanding, and can de jure or de facto exclude women agricultural workers (who are concentrated in seasonal and temporary labour force). Furthermore, maternity leave is often considerably shorter than internationally recognized standards (Fiji and Kenya). In addition, while in some countries maternity leave is fully paid, in others it is unpaid or it covers only a limited portion of the full wage (Fiji, Italy, South Africa and Tunisia). Finally, where maternity benefits are wholly or in part paid by the employer (Burkina Faso, Fiji, India, Mexico and the Philippines), women's access to employment is hindered by their higher labour costs.

Finally, women's labour rights are severely limited by the lack of implementation of labour legislation. For example, while most countries state the equal pay principle, gender pay gaps are reported for most of the examined countries. Pregnancy testing and even sterilisation practices have been documented in some countries (e.g. Mexico and Brazil). In plantations, women often work without contract on a daily and piece-work basis (as documented e.g. for Mexico, Brazil and South Africa). This deprives them of the protection accorded by labour law. In other cases, employment contracts are signed by the household head, and women provide labour as family members of employees; in these cases, wages are paid to the household head with regard to the global labour provided by the household (e.g. as documented for Tunisia and South Africa). More generally, a gender division of labour in agricultural work, whereby women are concentrated in low-pay, temporary agricultural work, is widespread in most of the covered countries, although to very different degrees.

G-8 COUNTRIES ON AGRICULTURAL LABOUR

GENEVA, Switzerland—Brazil's president, Luiz Inacio Lula da Silva, Monday slammed the world's leading industrialized countries for pursuing protectionist policies while preaching the virtues of free trade.

"Sectors in which developing countries have much more competitiveness such as in agribusiness, the textile industry, the steel industry, among many other sectors, are subject to protectionist trade practices from the highly industrialized world," he said in a keynote speech to the International Labour Organization here.

Lula said the resistance of the developed countries to eliminate the billions in agricultural subsidies, and their arbitrary practices, shows that they "are totally lacking coherence with their own defense of free trade. This lack of coherence between speech discourse and practice provokes skepticism and mistrust."

Lula's remarks echoed views shared by many other developing countries.

Brazil has strong views on agriculture are very much shared by us, a WTO ambassador from a major agricultural exporting country, who spoke on condition of non-attribution, told United Press International.

"We cannot be passive and just gaze at the disparity that exists between the islands of wealth and oceans of poverty," said Lula, a former trade union leader and child labourer.

But his remarks, coming less than a day after he and 12 other developing world leaders met with G-8 leaders during their annual summit Sunday in the resort town of Evian—on the French side of Lake Geneva—are at variance with the position of the G-8.

The G-8 includes the United States, Japan, Canada, Germany, the United Kingdom, France, Italy and Russia.

The divergence of views reflects a growing North-South divide ahead of a crucial WTO ministerial meeting in Cancun, Mexico, in September to try to inject momentum in the troubled WTO Doha round talks which faces a Jan. 1, 2005, deadline.

The Indian prime minister, Atal Bihari Vajpayee, also pressed G-8 leaders Sunday on the need to remove trade distorting agricultural subsidies and to eliminate tariff and non-tariff barriers hindering developing poor countries exports, trade sources said.

Kofi Annan, U.N. secretary-general, the chief of the World Trade Organization, Supachai Panitchpakdi, and the heads of the World Bank and the International Monetary Fund also attended.

On June 1, the U.N. chief told a G-8 leaders' session that developing countries "need better access to global markets, which means a Doha round that lowers agricultural subsidies and brings down barriers to imports from poorer countries."

Ahead of the Evian summit, Annan in a letter to G-8 leaders argued that decisions on trade would be especially crucial for the developing world.

"It is through trade that the peoples of developing countries can hope to share in, and contribute to, the general prosperity of mankind," Annan said.

Meanwhile, in a statement on trade issued Monday at Evian, the G-8 leaders stressed that they will provide leadership in the talks "so that improved access to markets for all WTO members is realized, particularly for the poorest."

'Energizing the Global Economy'

They also pointed out the global talks are central to the G-8's approach to "energizing the global economy," increasing development and eradicating poverty.

10

Consumer Price Index Numbers for Agricultural and Rural Labourers

INTRODUCTION

The Labour Bureau has been compiling CPI Numbers for Agricultural Labourers since September, 1964. The existing series of CPI Numbers for : (i) Agricultural and (ii) Rural Labourers (base 1986-87=100) replaced the earlier series on base 1960-61=100 w.e.f. November, 1995. For compilation of these index numbers, the Field Operations Division (FOD) of the National Sample Survey Organisation (NSSO) collects regularly every month the data on prices from 600 sample villages selected from 20 States. Consumer expenditure data collected by the NSSO during 38th round of NSS (1983) formed the basis of the weighting diagrams for the series. The methodology approved by the Technical Advisory Committee on Statistics of Prices and Cost of Living (TAC on SPCL) is followed for compilation of CPI-AL/RL for 20 States and all-India separatley for : (i) Agricultural, and (ii) Rural Labourers and these indices are released on monthly basis by 20th of every succeeding month.

CONCEPTS AND DEFINITIONS

Rural Labour Households

Rural Labour Households are those households whose income, during the last 365 days, was more from wage paid manual labour (agricultural and/or non-agricultural) than either from paid non-manual employment or from self-employment.

Agricultural Labour Households

The rural labour households, who derive 50 per cent or more of their total income from wage paid manual labour in agricultural activities, are treated as agricultural labour households.

Rural Labourer

Rural Labourer is a person who does manual work in rural areas in agricultural and/or non-agricultural occupation in return for wages in cash or kind, or partly in cash and partly in kind.

Agricultural Labourer

A person is treated as an agricultural labourer, if he/she follows one or more of the following agricultural occupations in the capacity of a labourer on hire, whether paid in cash or kind or partly in cash and partly in kind:

(a) Farming including cultivation, growing and harvesting of any agricultural commodity;
(b) production, cultivation, growing and harvesting of any horticultural commodity;
(c) dairy farming;
(d) raising of livestock, bee-keeping or poultry farming; and
(e) any practice performed on a farm as incidental to or in conjunction with the farm operations (including any forestry or timbering operations and the

preparation for market and delivery to storage or to market or to carriage for transportation of farm products).

The manual work in fisheries was, however, excluded from the category of agricultural labourer. Further, carriage for transportation coming under the category (e) above, referred only to the first stage of transportation from farm to the first stage of disposal.

SCOPE AND COVERAGE

The Consumer Price Index Numbers for Agricultural Labour cover the households of agricultural labourers and the Consumer Price Index Numbers for Rural Labour cover the households of rural labourers (including agricultural labourers). The rural retail prices utilised in these two series of index numbers are the same, but the weighting diagrams used for compiling these indices are different. As regards coverage of States, a Technical Working Group on Rural Retail Prices was constituted in 1974 under the Chairmanship of the Director, CSO which recommended 20 States, for which separate series of CPI Numbers were to be maintained. For Union Territories, the Group felt that an attempt to compile separate series would increase the field work considerably without commensurate increase in utility. Therefore, the Group recommended that the Union Territories could use the index series of the adjoining States.

The State-wise sample sizes were decided by studying the variations in village wise prices in the old series of index numbers and keeping in view the need for constructing State-wise CPI Numbers for : (i) Agricultural, and (ii) Rural Labourers. The sample size for a State was allocated to the regions within it in proportion to their rural population and in multiples of three. A number of regions were formed within a State, according to similarity of agro-climatic conditions and density of population. Within each region, strata were formed by treating each district as a separate stratum. However, the smaller districts, within a particular NSS region, were merged to form a separate stratum. The allocation to a stratum was generally 3 villages, except in case of 13 Strata (one each in

Assam, Karnataka, Manipur and Meghalaya; two each in Jammu & Kashmir, Kerala and West Bengal and three in Orissa) which had six villages each.

WEIGHTING DIAGRAM

The consumer expenditure data collected by the NSSO, in its 38th Round in 1983, was used for deriving weighting diagrams for all the 20 States and All-India, separately for Agricultural Labourers and Rural Labourers. The average budget derived from the household consumer expenditure survey recorded expenditure on a large number of goods and services. These items of expenditure were divided into two categories viz., (i) non-consumption out-go including unpriceable items, and (ii) consumption expenditure. Only consumption expenditure was taken into account for derivation of weighting diagrams. Expenditure on items like rent and repair of residential buildings or land, direct taxes, priest services (including ceremonial expenditure) and precious jewellery or ornaments (like gold, silver, pearl, etc.) was treated as non-consumption expenditure and thus excluded from the average budget. The expenditure on rest of the items (including the durable goods) in the average budget was considered as consumption expenditure.

In order to facilitate derivation of weights at item/sub-group/group level and calculation of the index, the groups and sub-groups were formed, to the extent possible, on the pattern of existing series of CPI Numbers for industrial workers on base 1982=100 and in pursuance of the recommendations of the TAC on SPCL. The expenditure was categorised into following groups :

(i) Food;
(ii) Pan, Supari, Tobacco and Intoxicants;
(iii) Fuel and Light;
(iv) Clothing, Bedding and Footwear; and
(v) Miscellaneous.

As it was not feasible to include all the items of expenditure in the index basket because of their un-manageability due to large number or existence of items which can not be priced easily or non-availability of regular price data for some of the items, in order to meet such an exigency, recourse to imputation of expenditure was undertaken.

Imputations were done at four levels, viz.: (i) item level, (ii) section level, (iii) sub-group level, and (iv) group level. Item level imputation consisted of straight addition of expenditure of one or more unpriced items to a priced item within the same sub-group/group. Section level imputation consisted of proportionate distribution of expenditure of one or more unpriced items over several priced items which formed a distinct section within a sub-group/group. In such cases, the unpriced items were assumed to follow the combined price trend of more than one priced item within the same section. The remaining two levels of imputation consisted of proportionate distribution of expenditure of one or more unpriced items over all the priced items, included in the sub-group/group. In such cases, an excluded item could not be taken to follow the price trend of any specific priced item of the sub-group/group, nor the combined price trend of a section of included items. The implicit assumption behind the imputation was that the price behaviour of the imputed item was the same as that of the item/section/sub-group/group in which its expenditure had been imputed.

After completion of expenditure imputation process, as envisaged above, the resultant expenditure on each item within a sub-group/group was expressed as percentage of the total expenditure on the sub-group/group to obtain item weights within a sub-group/group. Similarly, sub-group/group weights were derived by working out the percentage expenditure, including sub-group/group level imputations, on each sub-group/group as compared to total consumption expenditure. This provided the weighting diagram for compilation of the Consumer Price Indices for (i) Agricultural and (ii) Rural Labourers of each State. From the State-level weighting diagrams, all-India weights have been derived.

SELECTION OF BASE YEAR

A number of factors are taken into account for determining the base year of the series. A period affected by developments of serious nature such as war, floods, draught etc. is not adopted as the base year because it cannot be treated as a normal year.

CAUSES

(i) absence of adequate social support infrastructure at the level of the village and district
(ii) uncertainty of agricultural enterprise in India
(iii) indebtedness of farmers
(iv) rising costs of cultivation
(v) plummeting prices of farm commodities
(vi) lack of credit availability for small farmers
(vii) relative absence of irrigation facilities
(viii) repeated crop failures

REMEDIES

(i) government to actually implement the various money-lending Acts that already exist to prevent the alienation of the farmers land-holding
(ii) to make the crop Insurance Scheme more farmer friendly, with lower premia and less red-tape
(iii) renewal of the land's biodiversity to ensure the health of land and enable the farmer to cope with market ups and downs
(iv) better health facilities in the locality since expenditure on health has been one of the most important financial drain in the village
(v) better education facilities at school level in the villages to enable better coping with a more technologically oriented agriculture
(vi) quality checks on agricultural inputs like seeds, fertilizers and pesticides to prevent cheating of the farmer by unscruplous suppliers of industrial inputs for agriculture
(vii) reliable agricultural advisories for farmers on farm related practises
(viii) better access to markets for agricultural produce to get higher rates for farm produce

CURRENT SITUATION IN VIDHARBA

Vidarbha was in the media spotlight for a spate of farmer suicides in recent years ostensibly because of the falling Minimum Support Price for cotton. The problem is complex and root causes include lopsided policies of the World Trade Organisation and developed nations' subsidies to their cotton farmers which make Vidarbha's cotton uncompetitive in world markets. Consequently Vidarbha is plagued by high rates of school drop outs, penniless widows left in the wake of suicides, loan sharks and exploitation of the vulnerable groups.

The Indian government had promised to increase the minimum rate for cotton by approximately Rs. 100 ($2) but reneged on its promise by reducing the Minimum Support Price further. This resulted in more suicides as farmers were ashamed to default on debt payments to loan sharks. "In 2006, 1,044 suicides were reported in Vidarbha alone—that's one suicide every eight hours."

On 1 July 2006 the Prime Minister of India Manmohan Singh announced a Rs. 3,750-crore (Rupees 37.5 billion) relief package for Vidharbha. The package should help farmers in six districts of the region. However not everybody is convinced that the aid is getting through to where it is needed. Activists covering the region feel that a lot more needs to be done. A fortnight after the PM's package was announced, journalist P Sainath wrote the following article in the Hindu criticising the package and saying that it was destined to fail.

In April 2007 an NGO named Green Earth Social Development Consulting brought out a report after doing an audit of the state and central government relief packages in Vidarbha. The report's conclusions were:

(i) Farmers' demands were not taken into count while preparing the relief package. Neither were civil society organisations, local government bodies, panchayats etc consulted.

(ii) The relief packages were mostly amalgamations of exiting schemes. Apart from the farmer helpline and the direct financial assistance, there was scarcely anything new being offered. Pumping extra funds into additional schemes shows that no new idea was applied to solve a situation where existing measures had obviously failed.

(iii) The farmer helpline did not give any substantial help to farmers

(iv) The basis for selection of beneficiaries under the assistance scheme was not well-defined. Also, type of assistance to be given led to problems like a farmer needing a pair of bullocks getting a pump set and vice versa (or a farmer who has no access to water sources being given pump sets)

(v) Awareness regarding the package was also pretty low.

INDEX